目录

CONTENTS

HONG KONG & TAIWAN

MODERN VILLA

LEADING DESIGN

港台豪宅设计典范

高迪国际出版有限公司 编

杨显艳 张春艳 荣亮 译

大连理工大学出版社

图书在版编目(CIP)数据

港台豪宅设计典范：英汉对照 / 高迪国际出版有限公司编；杨显艳，张春艳，荣亮译. — 大连：大连理工大学出版社，2014.7

ISBN 978-7-5611-9181-1

Ⅰ.①港… Ⅱ.①高… ②杨… ③张… ④荣… Ⅲ.①住宅—室内装饰设计—英、汉 Ⅳ.①TU241

中国版本图书馆CIP数据核字（2014）第108943号

出版发行：大连理工大学出版社
（地址：大连市软件园路80号　邮编：116023）
印　　刷：上海锦良印刷厂
幅面尺寸：240mm × 290mm
印　　张：20
插　　页：4
出版时间：2014年8月第1版
印刷时间：2014年8月第1次印刷
责任编辑：初　蕾
责任校对：王丹丹
封面设计：高迪国际

ISBN 978-7-5611-9181-1
定　　价：320.00元

电 话：0411-84708842
传 真：0411-84701466
邮 购：0411-84708943
E-mail：designbookdutp@gmail.com
URL：http:// www.dutp.cn

如有质量问题请联系出版中心：（0411）84709043　84709246

CENTER STAGE

聚贤居

设计公司
Danny Cheng Interiors Ltd.

设 计 师
郑炳坤

项目地点
中国香港

项目面积
233m²

The project is located in Central Hong Kong, with elaborate design and skillful application of colors and thus a fashionable residence in the busy downtown has been built up.

An open kitchen combined with a half-height bar cabinet brings out atmosphere of interaction, which enables people to have dinner here and increases the intimacy between the host and guests. Light grey marble-floor and clear mirror walls not only enlarge the space visually but also create a stylish and fashionable feeling. In the sitting room, there hasn't been installed with a TV but a wall is served as the screen for projection, therefore, the spatial stereovision has been enhanced and visual effects have been enriched. A long purple rug adds a soft ambience to the living room and it highlights the charming perspective of city life.

A colorful suspended stairway leads to the layer of bedroom, and bright colors connect the two levels, adding vitality to the space. The black stainless steel door beside the stairway covers the storage and it can be opened when needed, which sets the master and the guest bedroom apart and then a sense of spaciousness and penetrating is added to the space of the lower level. There is a comfortable bed and a massage bathtub in the master's bedroom without much decoration. The master can enjoy the view outside the window while taking a bath and it is enough to provide a relaxing and comfortable mood after a busy day's work. The white wardrobe behind the bed constitutes a cloakroom, which is a way of making full use of space. Stainless steel shelves and red circular sofa are decorated in the guest bedroom which is filled with an intoxicating atmosphere.

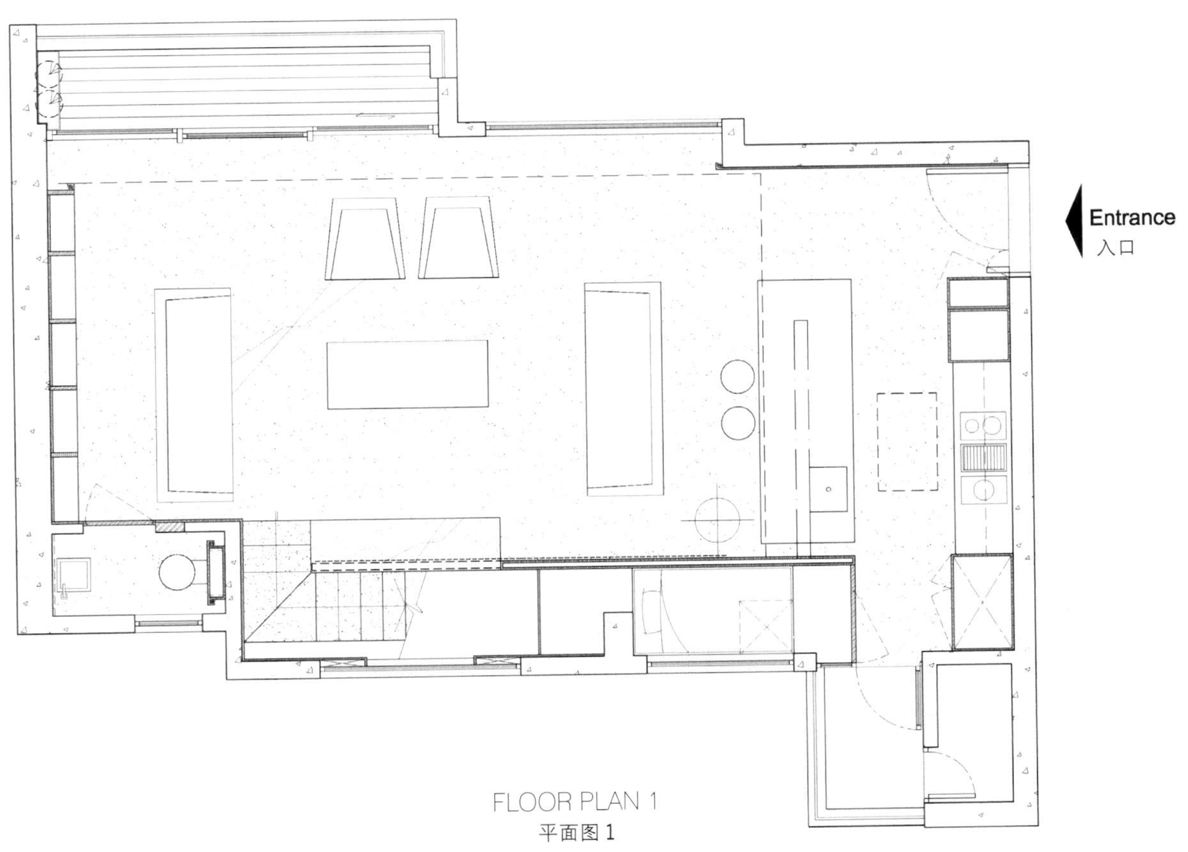

FLOOR PLAN 1
平面图 1

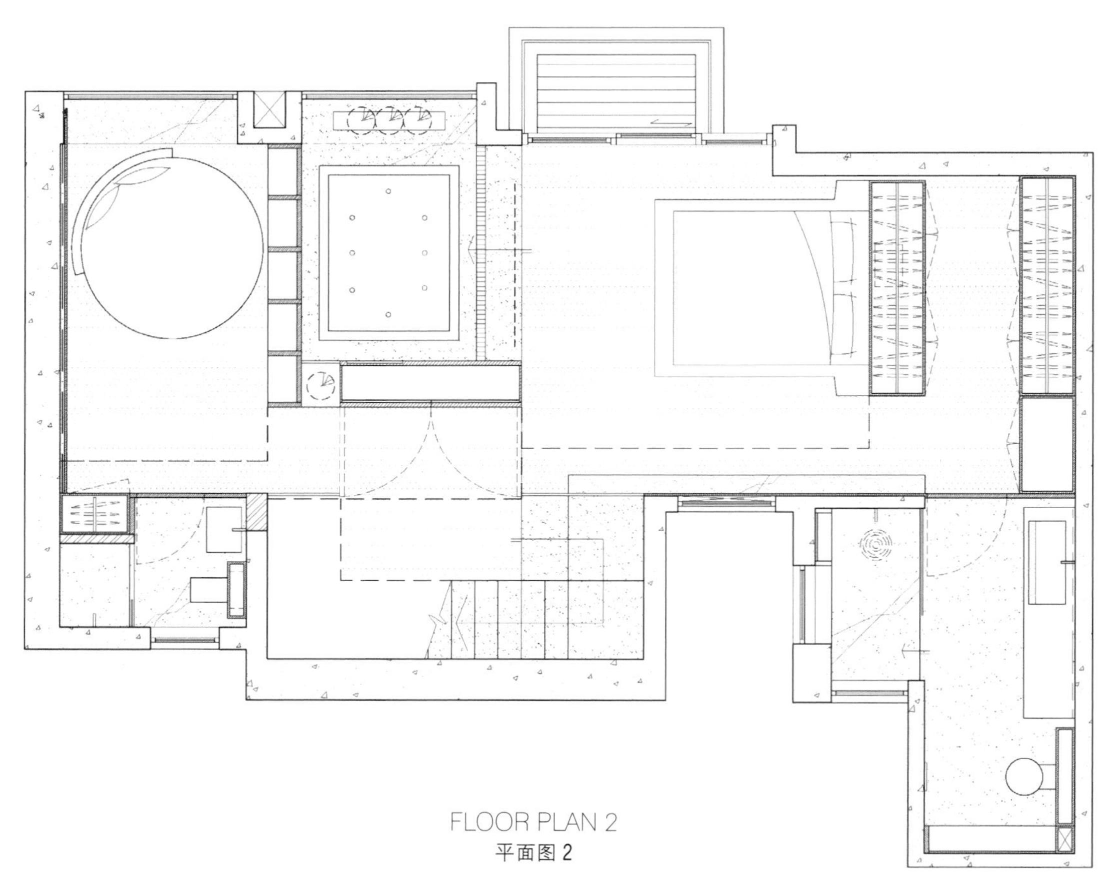

FLOOR PLAN 2
平面图 2

MAIDENS GREEN
PLAISTOW GREEN
WOKINGHAM
CROWTHORNE
HAWTHORN HILL
BRACKNELL
CANNON CORNER
BROOKSIDE
SUNNINGDALE
BRACKNELL (WILDRIDINGS)
BRACKNELL (GREAT HOLLANDS)

这个位于中环的单位，经过精心的设计及巧妙的颜色运用，在繁华闹市中打造了一个时尚的家居环境。

开放式的厨房配合半高的吧柜，客人可于吧柜上用餐，带出互动的气氛，消除了屋主与客人的阻隔。浅灰色的云石地面与清镜墙身，在增加空间感之余营造出时尚亮丽的感觉。客厅没有安装电视机，取而代之利用墙身作投影幕，提升了空间的层次感，丰富了视觉画面。紫色长形地毯为客厅添上柔和的情调，突显了城市迷人的一面。

彩色的悬浮楼梯带你进入睡房层，用色彩连接上下两层，为空间增添了生机。楼梯旁的黑色不锈钢掩门将储物空间掩藏起来，需要时可将门打开，区分出主人房与客房，楼梯下的空间及玻璃间则隔增添了空间感及通透性。主人房没有太多的装设，一张舒适的睡床及按摩缸，可令屋主于忙碌的生活中放松心情。一面浸浴，一面欣赏窗外景致，格外舒适惬意。床背后的白色衣柜构成了衣帽间，充分利用了空间。客房内的不锈钢层架配合红色圆形沙发，令房间充满醉人的气氛。

THE HARBOURSIDE

君临天下

设计公司
Danny Cheng Interiors Ltd.

设 计 师
郑炳坤

项目地点
中国香港

项目面积
155m^2

The impression of white is simple and fashionable and this color combined with an elaborated design fully conveys a modern sense.

White light panel and gray mirrors serve as the main materials in the kitchen and the design is to create a stylish, bright and clean cooking space. The kitchen door and the sitting room wall are carved into a unique concavo convex effect which adds fun to the wall. Meanwhile, the full-height shelves on one side of the dinning room are convenient for the owner to place things as well as enrich the colors in the dinning room. The long gray sofa is integrated with a black long rug perfectly. The TV is hanged on a column, of which the design is distinct from traditional ones and it enhances the modern impression. Sunlight irradiates into the sitting room through full-height windows beside the sofa and thus brings warmth to the room. In the sitting room, the desk with mirror surface near the window could enable the host to read or rest while enjoying the view outside the window. On the side of the sitting room is a folding bed, and a guest room will be formed once the big door in the sitting room is moved to the side. This design is very flexible to make more space available.

Wooden floors in the sitting room and dining room extend to the study and the master bedroom, the design is concise and expressing a sense of warmth. The open study is unconstraint for the master to read on the desk in the abundant sunlight beside the window, with the tear-dropped ceiling light together adding fun to the room. The light purple windowsill mattress and the bed frames increase the degree of emotion and sentiment as well as the sense of tenderness. The open door of the white wardrobe is carved with tower patterns which add a stylish atmosphere and it can hide the dresser, thus all the elements in the design are integrated together harmoniously and a comfortable and concise feeling is effectively conveyed.

This modern residence is very comfortable to live in, and provides a relaxing mood for the host in the busy city life.

CLOSE IT
CLOSE IT

白色让人感觉简洁时尚，配以设计师精心的设计，更加将现代感表露无遗。

厨房以白色光面板及灰镜为主要材料，打造出时尚亮丽且整洁的煮食空间。厨房敞门及客厅墙身以特别的方法雕刻出独一无二的凹凸效果，为墙身增添了趣味性。餐厅一边的全高层架，可方便屋主摆放东西，也为餐厅增添了色彩。灰色长形沙发与黑色长形地毯完美搭配在一起，电视机悬挂于圆柱上，有别于传统设计，提升了现代感。沙发旁落地玻璃窗将阳光引入，投射到客厅，为客厅增添了温暖。偏厅内靠窗的位置摆放了镜面书桌，可让屋主一边休息阅读，一边欣赏窗外景色。偏厅内收藏了一张折叠床，只要将客厅中的大敞门推至偏厅，就成了一间客房。设计灵活度大，使空间用处变多。

客厅、餐厅的木地板延伸至书房及主人房，简约中透出温暖感。开放式书房让人感觉更无拘无束，配合水滴形天花灯为房间增添趣味性。靠窗书桌让屋主在充足的光线下阅读。睡房内浅紫色的窗台坐垫及床架为睡房提升了情调，令空间更为柔和。白色的敞门衣柜上刻有铁塔图案，增添了时尚气息。敞门可将梳妆台掩藏起来，让整个空间相统一，带出舒适而简洁的感觉。

这个具时代感的居所让人感到舒适自在，于繁忙的都市中放松心情，给屋主一个无压力的家。

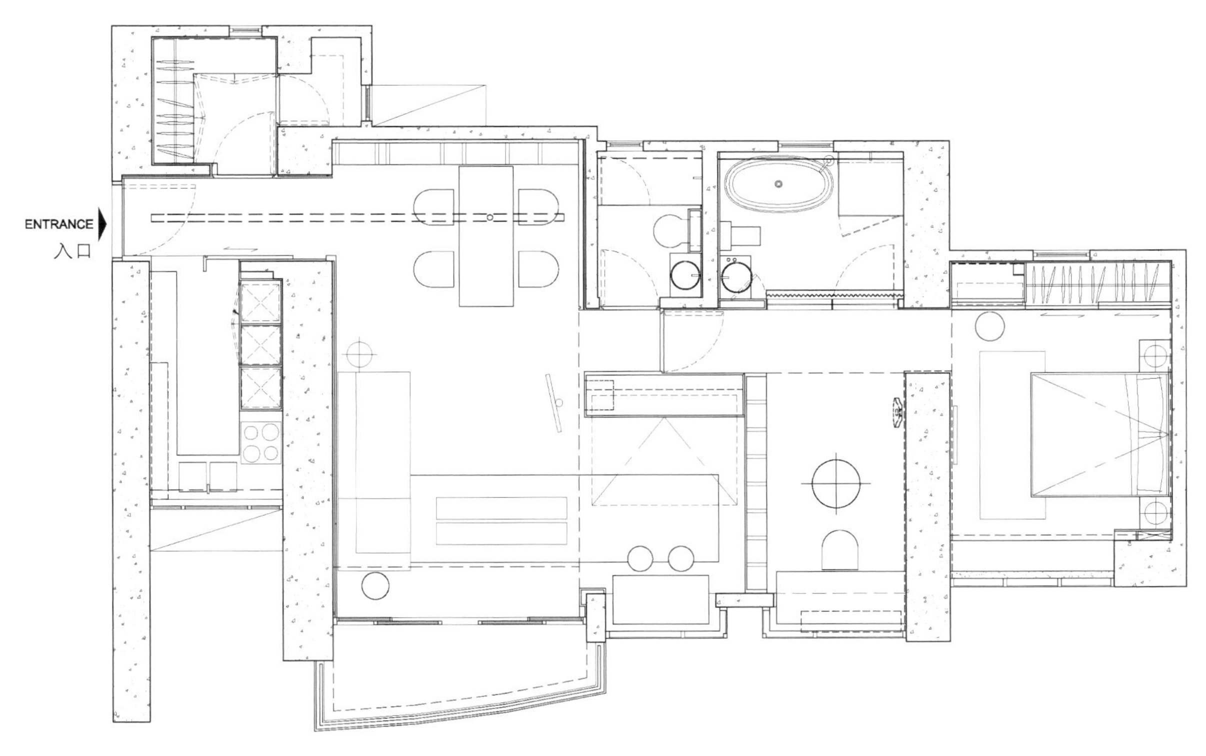

FLOOR PLAN
平面图

THE WESTMINSTER TERRACE

皇璧

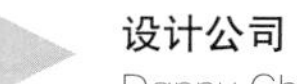

设计公司
Danny Cheng Interiors Ltd.

设 计 师
郑炳坤

项目地点
中国香港

项目面积
355m²

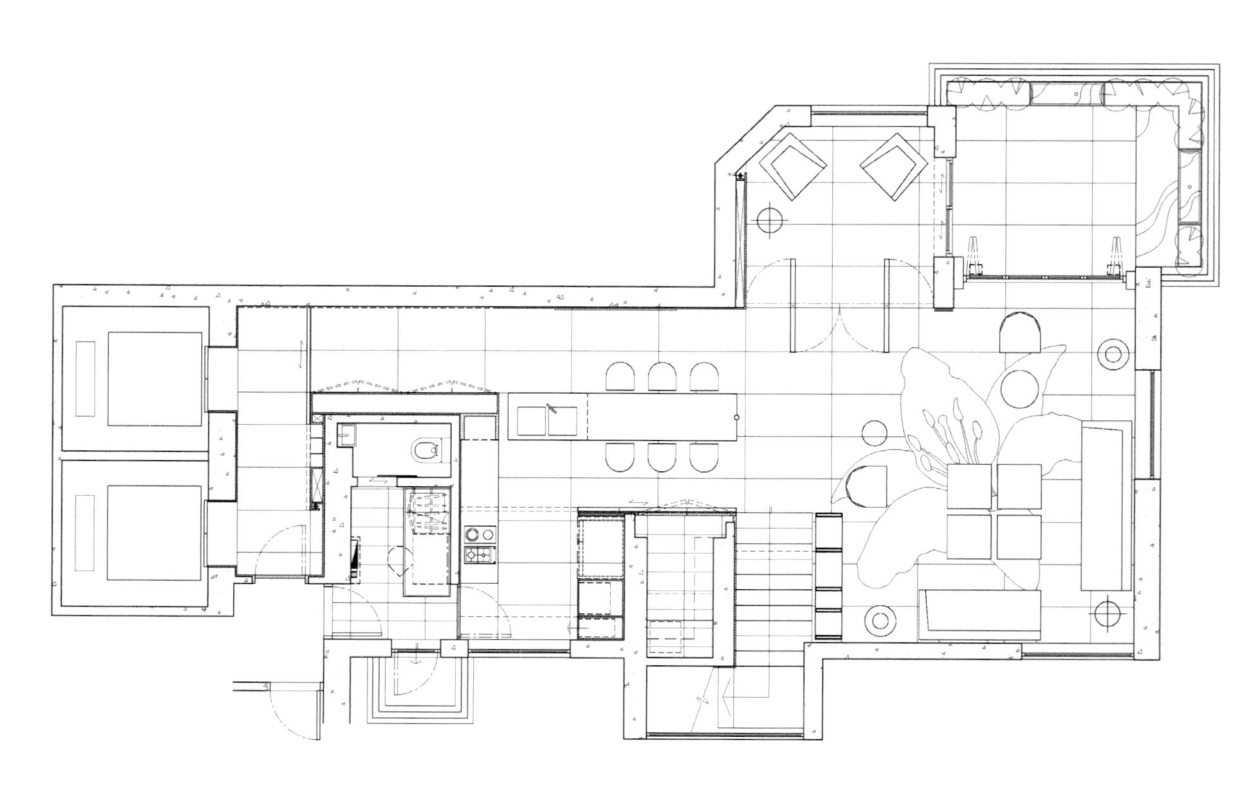

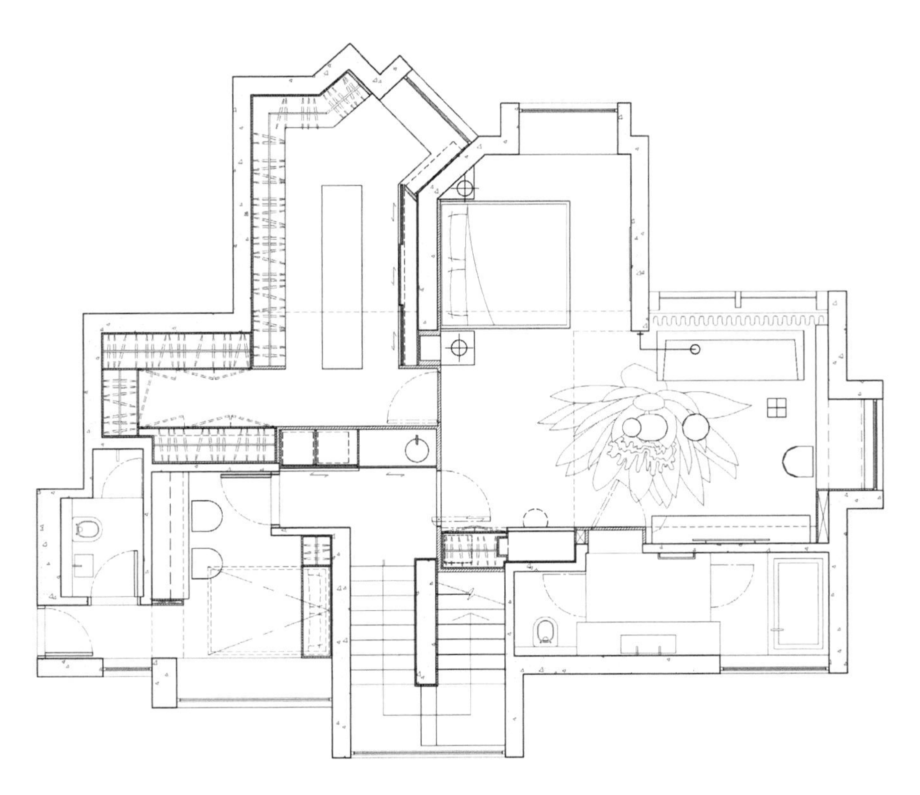

This duplex apartment has been furnished and decorated to highlight an extraordinary sense by the designer with a delicate design style.

The electrically operated gate welcomes you to this magnificent apartment. The mirror walls by the door not only extend space but also reflect the sight at the dining room, which adds something fun. The open kitchen is equipped with a long white table with a gemellion, open and interactive. Streak wooden revolving door zones the living room into a space with high privacy, thus becoming the highlight. High TV bench of steel-mirrors material provides space for storage, and upgrades the elegant feeling of the living room to project the advantage of its large base height as well. Marble used as the platform of living room and dinning room is extended into the open-air terrace, elegant and fashionable. A carpet with flower patterns matches well with the plants on the terrace and the murmuring of a brook, building a natural and imaginative space atmosphere. From the kitchen, the dining room to the living room, the whole space looks penetrating and integrative without the disruption of partition walls.

The stairs and the floors in the bedrooms are covered with wood floor to add the feeling of warmness. The master's room is colored with brown, conveying a sense of harmony. Upholstery is used for covering the bed to increase comfort. A small side hall provides the owner with enough room to get rested and relaxed. The carpet with flower patterns invigorates the master's room and echoes one another with the living room. The cloakroom, decorated with wooden leather exploits space to allow the owner to store as much as possible. A single glass-cabinet which can be used to exhibit such accessories as jewelries and watches is an excellent combination of beauty and practicability.

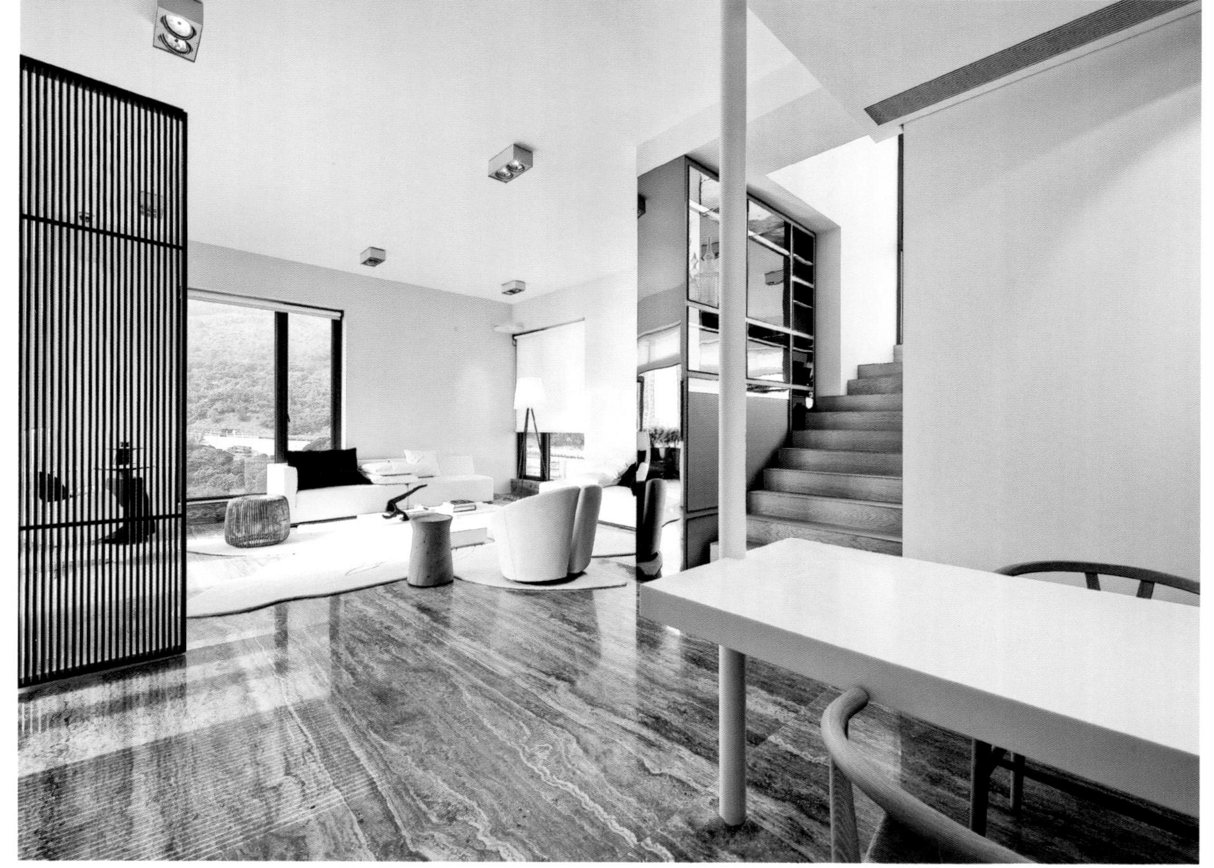

设计师以精致的设计突显出属于这个复式住宅单位的非一般感觉。

一道电动大门带你走进这个气派的单位，大门旁的镜墙身在增加空间感之余，亦反映了餐厅的景象，增添了趣味性。开放式的厨房配以设有洗手盆的长形白色餐桌，给人开放互动的感觉。木皮条子旋转门成为客厅的焦点，可将客厅划分出一个隐私度较高的空间。全高镜钢电视柜提供收纳功能的同时，也提升了客厅的格调，突显了楼底高的优势。以云石作为客厅和餐厅的地面，并将其伸延至户外露台，既高雅又时尚。客厅的花型图案地毯配合露台的植物、流水声，增添了空间的大自然气息，营造出写意的气氛。由厨房、餐厅到客厅，没有间隔墙身，整个空间富有通透感及连贯性。

楼梯及睡房层铺设了木地板，增添了休息空间的温暖感。主人房以棕色为主色调，带出和谐的气氛。床背以扪布作墙身，加添舒适的感觉。偏厅提供了充足的空间以便休息及放松心情。花型地毯为主人房增添了生气，并与客厅互相呼应。以木皮作主要材料的衣帽间，充分地利用空间，让屋主可摆放大量衣物，配合玻璃面的独立柜，可陈列首饰、手表等饰物，美观与实用性并重。

HEWEN TIANSHAN UNIT

何文田山单位

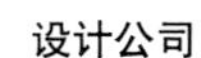

设计公司
Danny Cheng Interiors Ltd.

设 计 师
郑炳坤

项目地点
中国香港

项目面积
144m^2

主要材料
云石、木材

This remodeled residence is given a good transparency and spaciousness by using open kitchen and stripe screen to divide the space.

At the doorway, one side of the walls is surfaced with a mirror which can escalate the sense of space as well as ensuring the owner's neat appearance before going out. The other side is full of tall cabinets to utilize every corner to enlarge storage space. The mobile stripe-screen divides the space into living room and dining room, which makes a clear distinction between space function and creates a sense of penetrating and brightness. The living room is adjacent to the open-air terrace, which not only brings sunshine into the room but also enlarges the vision and creates an enjoyable atmosphere. The lights are put in troffers and hidden corners in order not to spoil the unity of space. A hidden troffer extends from the back of the sofa to the walls of the open-air terrace, creating a visual effect of wholeness. Besides promoting the sense of spaciousness, the open kitchen also forms interactive surroundings. Transparent chairs make the whole space more penetrating and make the dining table look like a suspending one, which brings more amusing elements into the whole space.

The suspending stairs wind through the two storeys. The space under the staircase is for children's playing. The door of the up-level's study is designed into an extremely large one, which can not only expand your vision, but also avoid the sense of narrowness when opening all the way. The desk by the window has a good natural lighting, which makes you enjoy the outside scenery and take a full relaxation. L-shaped sofa in dark grey is suitable for reading and taking a break. Children's room is spacious enough for storage. The bedroom continues the concise tone of the whole design with cozy large bed and wooden floor offering warmness.

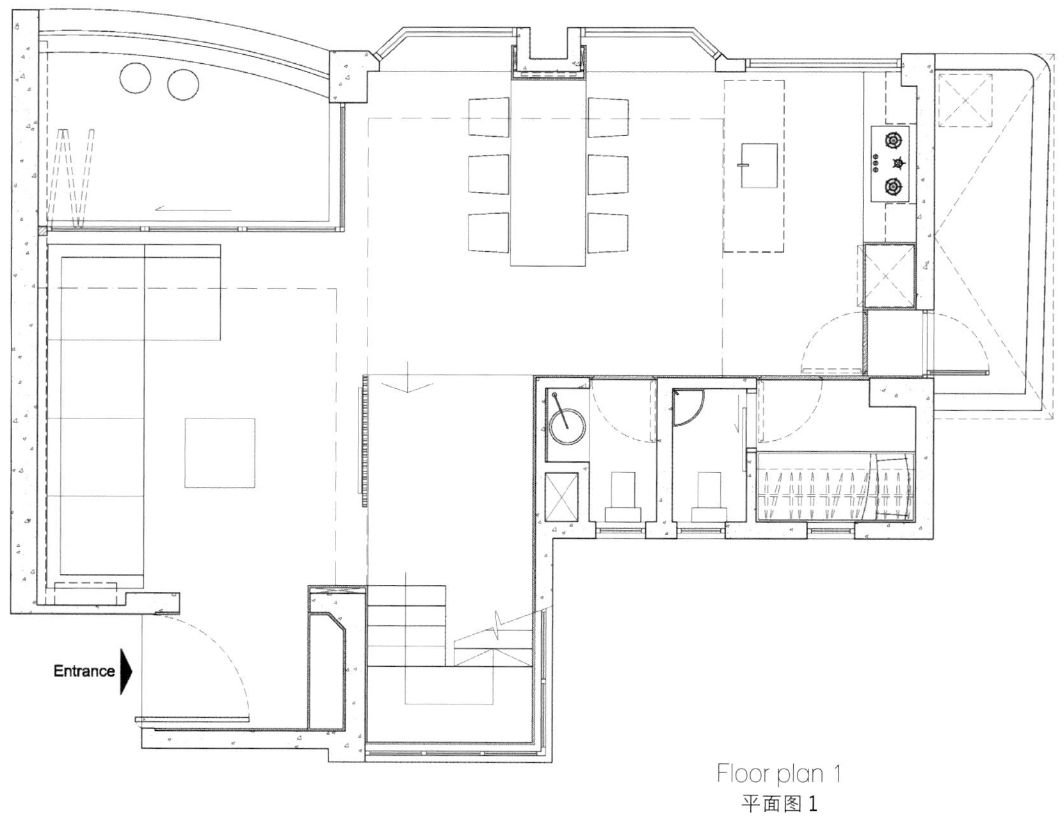

Floor plan 1
平面图 1

这个重新规划后的家居空间，利用开放式厨房及条子屏风来区隔空间，巧妙地提升了透明度及空间感。

门口一边的墙上铺贴了镜片，在提升空间感之余，也可让屋主在出门前稍作整理。另一边墙由全高柜子组成，利用每个角落扩大储物空间。以活动条子屏风区分客厅、餐厅，区域分明，通透性强。客厅连接着露台，不但为客厅引入了自然光，还扩宽了视野，营造出写意的气氛。利用灯槽及暗藏位置安排灯光，令整体气氛更加和谐融洽。沙发背的暗藏灯槽延伸至户外露台墙身，使视觉效果更统一。开放式的厨房除了增强空间感之外，还可以创造互动的环境。透明餐椅加强了视觉上的通透感，也突显了餐桌悬浮的效果，倍添趣味。

悬浮楼梯可增加上下两层的贯穿性，楼梯下的空间是日后小朋友玩乐的地方。上层书房的房门很大，当门完全打开时，能在视野上拓宽空间，不会显得局促窄逼。书桌靠窗既可自然采光，又可欣赏窗外景致，放松心情。深灰色的L形沙发是休憩阅读的好地方。小朋友房提供了足够的空间，可让屋主集中储物。睡房延续了简洁的主调，舒适的大床及木地板，给人温暖的感觉。

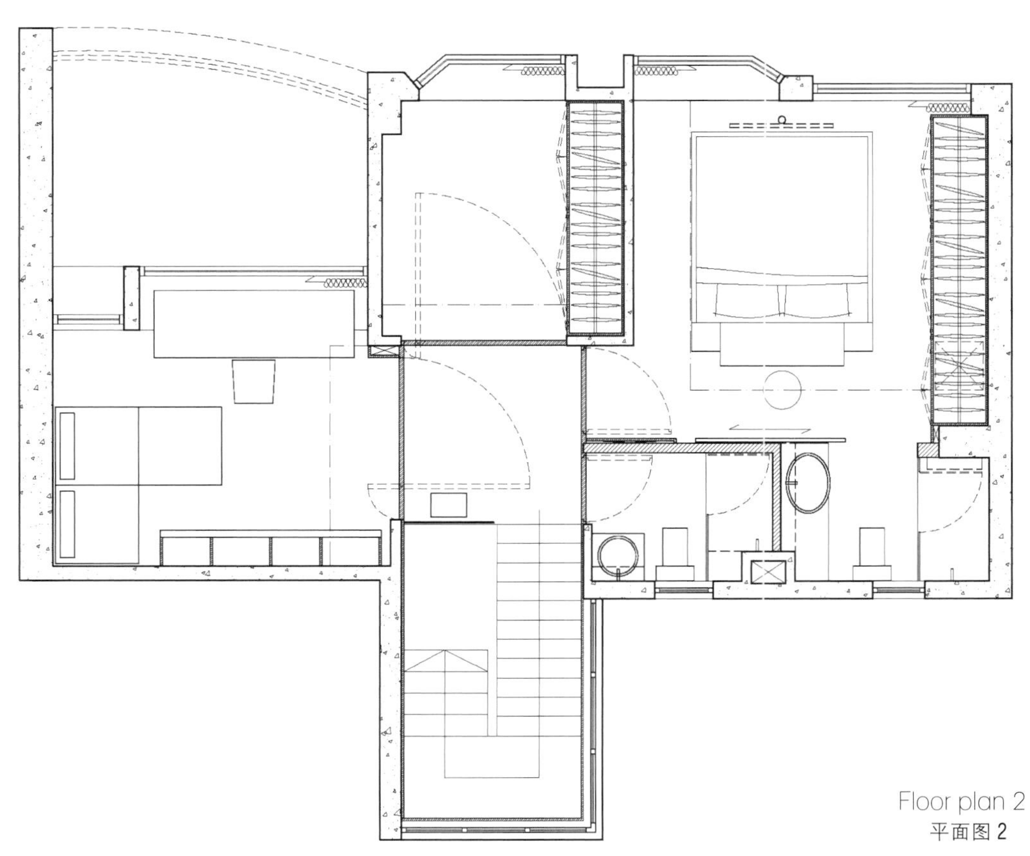

Floor plan 2
平面图 2

ELLE
DECORATION

GOLF

MR LI'S RESIDENCE – WATER PARK IN DANSHUI

淡水水公园李宅

设计公司
界阳 & 大司室内设计

设 计 师
马健凯

项目地点
中国台湾

项目面积
165m²

主要材料
硅藻土、卡拉拉白大理石、镜面不锈钢、亚克力、铁件、灰镜、人造石、灰玻、烤漆玻璃、水晶灯

In terms of designing aesthetics, invisible cabinets are used for integrating the storing function of diffierent time period; passing through the main door, the vertically entered sunlight, mirrors and the one piece stone extend almost 8 meters against the wall when they reach the window and "fold" into two-staged break over angles, making the surface tri-dimensional. The beam in the hall way, the horizontal lines first met the sky are adorned and beautified, and then all the static and dynamic elements integrated into the light atmosphere of the transparent screen.

The furniture echoes to each other revealing the client's taste, and is one of the highlights in front of the concise background, such as the irregular side coffee table which breaks the boundary of the study, it looks like the diamond break over angel with an eye-catching style and small objects can be put in it, which breaks the stereotype of common frame normally without the capacity of storing; while the combination of irregular theme chairs with uniquely customized dining table not only creates the focus of each area but also applies the general designing elements, integrates all the focuses in every view into a tremendous living space.

Though the spacious folded door is separated from the study, it doesn't hinder the extension of the doodle wall behind it. Real and the illusionary interlaced together, occasionally penetrating and occasionally obscured by the painted glass, the interesting blank canvas meets the creativity and imagination of children. The kid's room is based on the cartoon character Mickey, full of a distinct and interesting theme. From the delicate and small hardware to ceilings, bed frames, bed backplane, all the things combine together to create a festive atmosphere of Disneyland.

设计美学的呈现上，室内皆以暗柜处理来整合各季节的置物功能，开启大门，纵向迎来明亮光感，镜面、石材一体成型的概念在立面上延展约8米，临窗之处“折”出二阶段导角，将平面加以立体化。玄关入口处的大梁，在横向初遇天际的人文线条被修饰美化，一切动静交融于透明屏风的光感氛围中。

相互呼应并彰显品位的家具，是简约背景前的重心之一。如打破书房界线延伸而出的不规则3D边几，钻石导角切割的抢眼外型内，可放置物件，超乎了人们对框架没有储物功能的一般想象；而衬托不规则主题的造型椅、独一无二的订制餐桌椅组合，不但塑造了每个场域的焦点，更以概念相通的设计语汇，呼应并统合了各个视角内的精彩。

开阔折门虽独立出书房空间，却不阻碍后方涂鸦墙的延伸，虚实相间的面，偶尔穿透、偶尔以烤漆玻璃遮蔽，趣味十足的空白大画布，更满足了小孩天马行空的想象力与创造力。小孩房以Mickey卡通人物为题材，童趣鲜明的主题式空间，从精致细小的五金到天花板、床架、床背板，无处不为小主人营造迪士尼乐园的欢乐气氛。

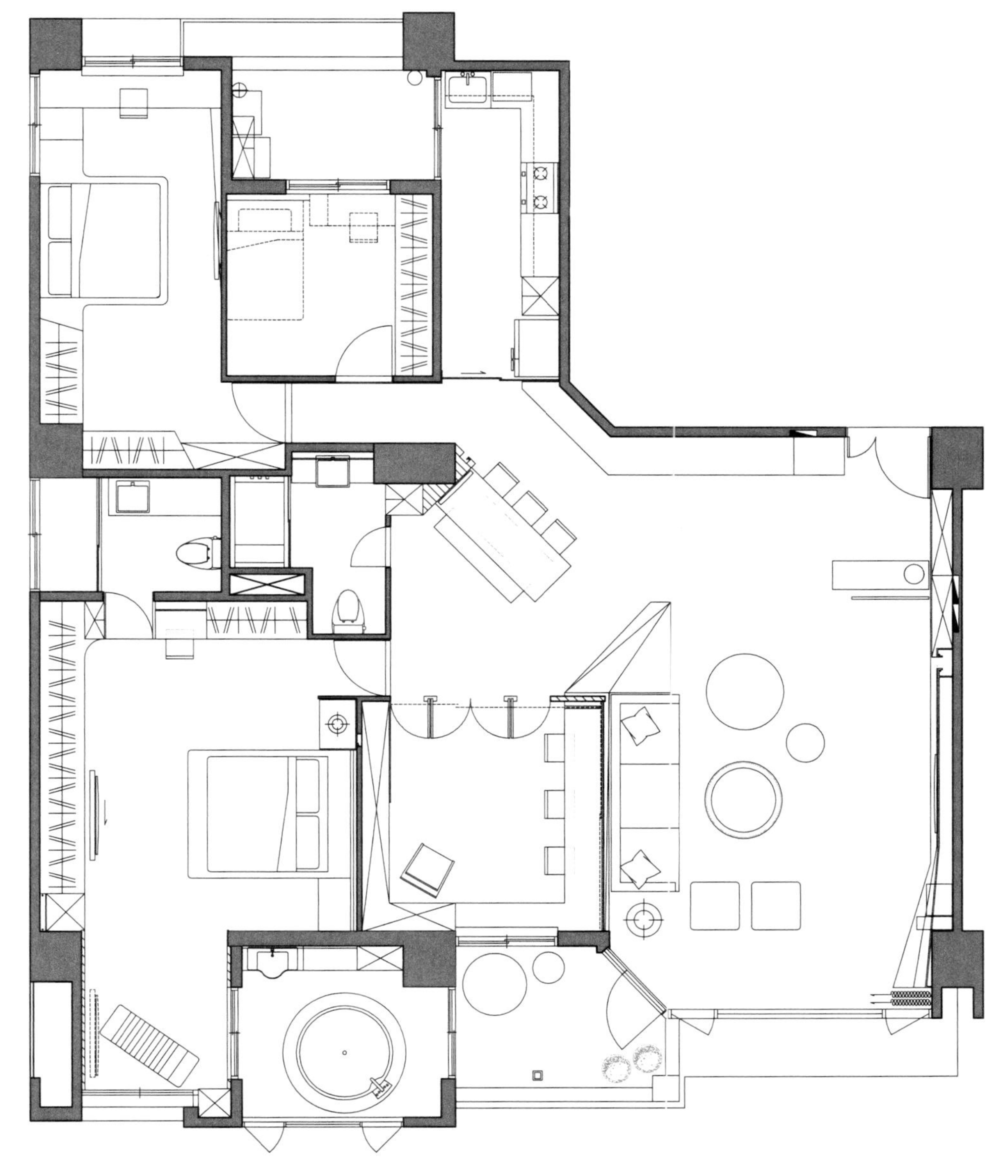

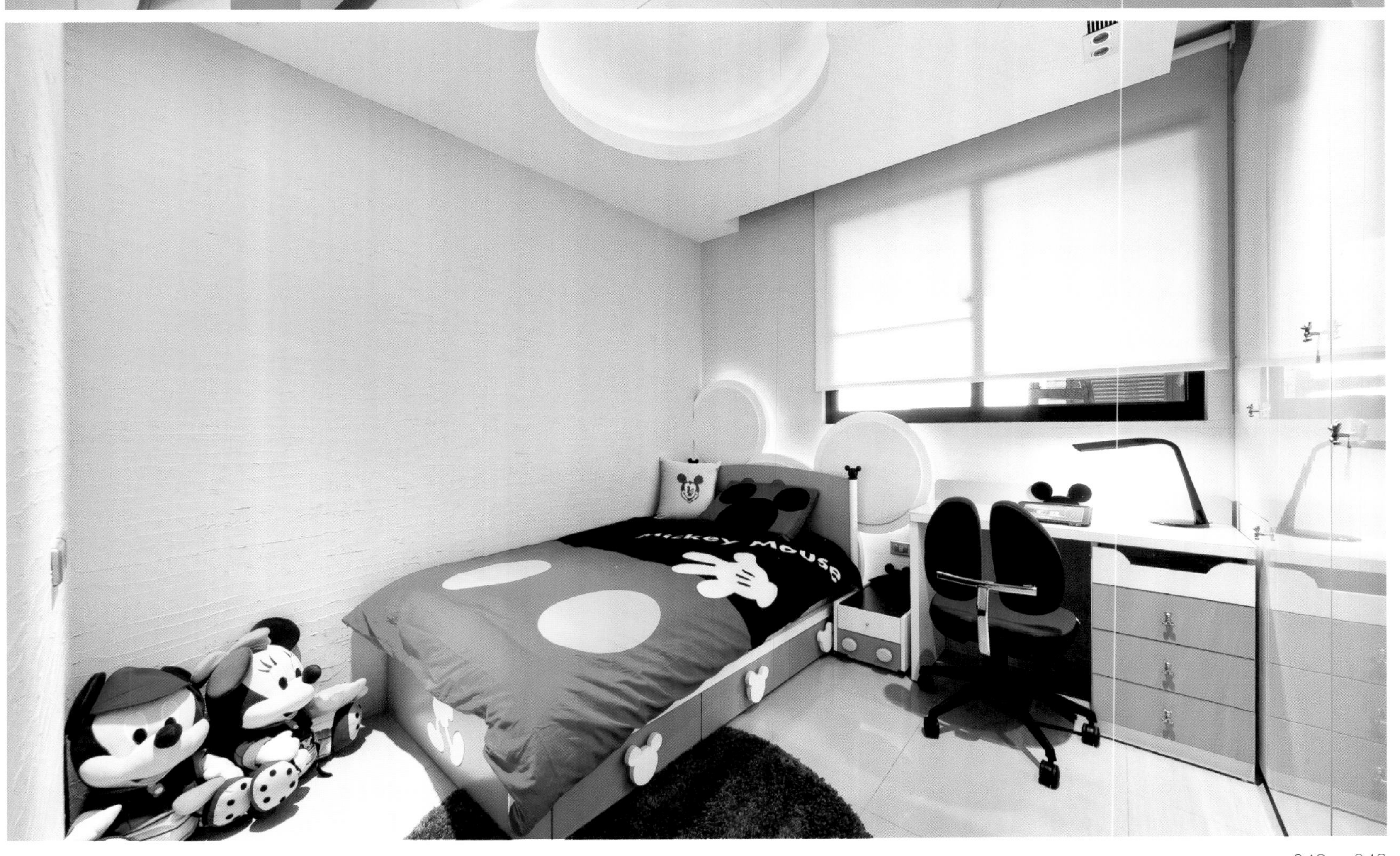

MR CHEN'S RESIDENCE IN TAOYUAN

桃园大观陈宅

设计公司
界阳 & 大司室内设计

设 计 师
马健凯

项目地点
中国台湾

项目面积
53m²

主要材料
镜面不锈钢、钢琴烤漆、进口壁纸、亚克力、LED

The three levels of arc TV wall extend to the gradually narrowed central part which is different in the depth of angles. The three levels link up to the ceiling lines, leading its inhabitants roaming with ease. The creative linear motion extends to the one piece desk in study, starting from the teapoy in living room, quickly rises, turns and extends, the design smoothly defines the boundaries of living room and study. And in terms of the lights atmosphere of the two spaces, special lighting equipment refracts a cool color light beam scattered on the central facade, which becomes a unique adornment symbol.

The high-quality youthful bedchamber and the milky white light box on one side of the entrance display both the husband and wife's collection. Half-height video wall is divided into triangle shapes. Black and white colors filled in contrast to present a strong visual intensity and the role of stainless steel frame adds quality of texture. The theme of random arc levels is reserved in furniture design. Once the original stainless steel is turned, beautiful light waves bloom in curved surface which diffuses the texture of material masculine.

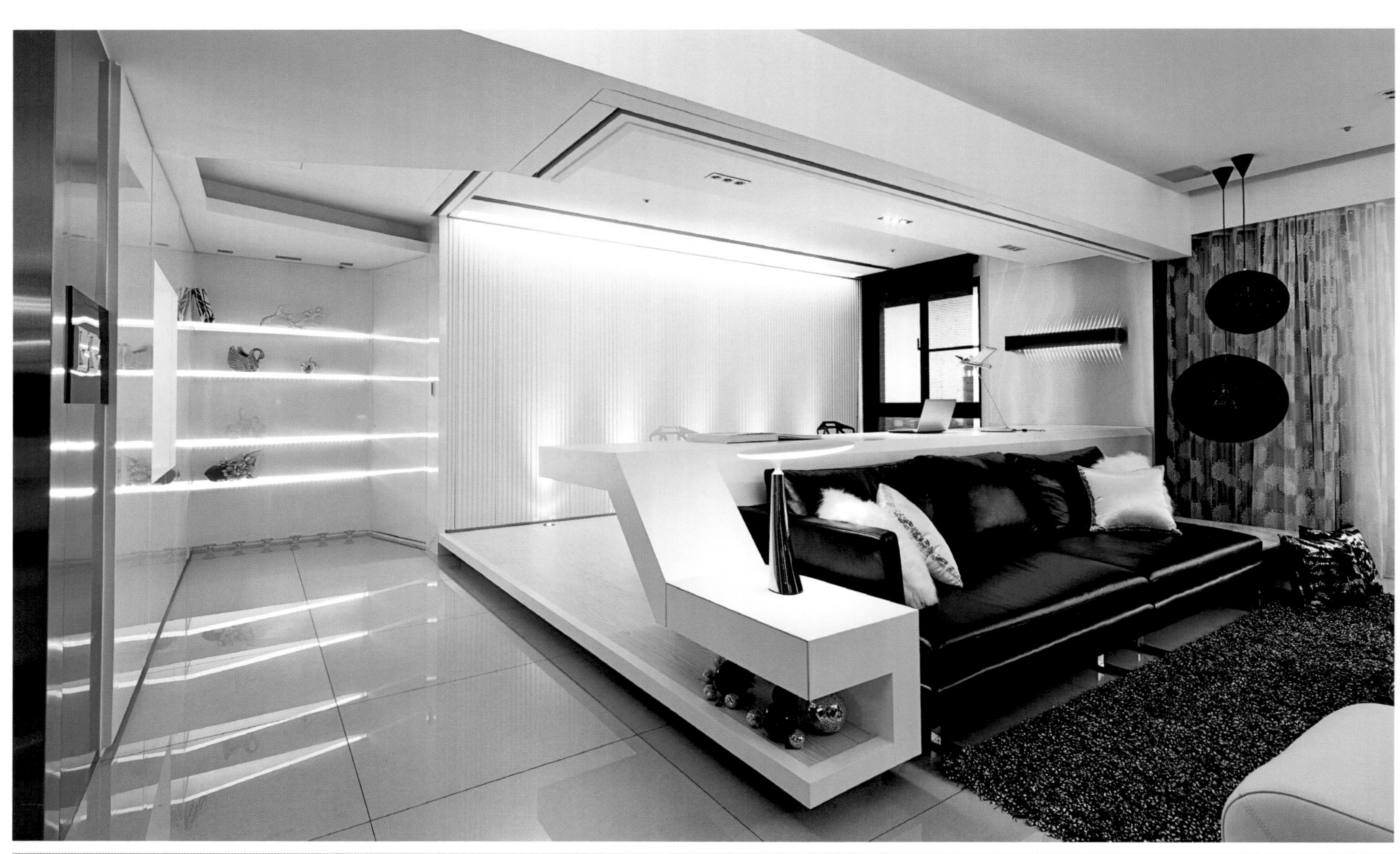

CHANEL

GUCCI
CHANEL

三层的弧形电视墙，深度向中央渐窄，形成差异化的角度延展。三个段落向上连结天花线条，轻松带出居住者的漫游脚步。创意的线性动态，延续至书房一体成型的书桌，以客厅茶几为起点，一气呵成地利落上扬、转折、延展，流畅地定义出客厅、书房的分界线，而在两个空间的灯光氛围上，特制的灯光设备，折射出的冷色光束，散落于中央立面，成为独特的装饰符号。

年轻化的精品主卧房，入口一旁的乳白色灯箱，高调地展示着夫妻俩的收藏品。半高影音墙的表面分割出三角块状，黑白填色对比出视觉强度，并框饰了不锈钢，平添了一份质感。自由随意的弧形层次，在家具设计中延续了主题。不锈钢的创意扭转在弯曲面绽放出美丽的波光，柔化了材质的阳刚味。

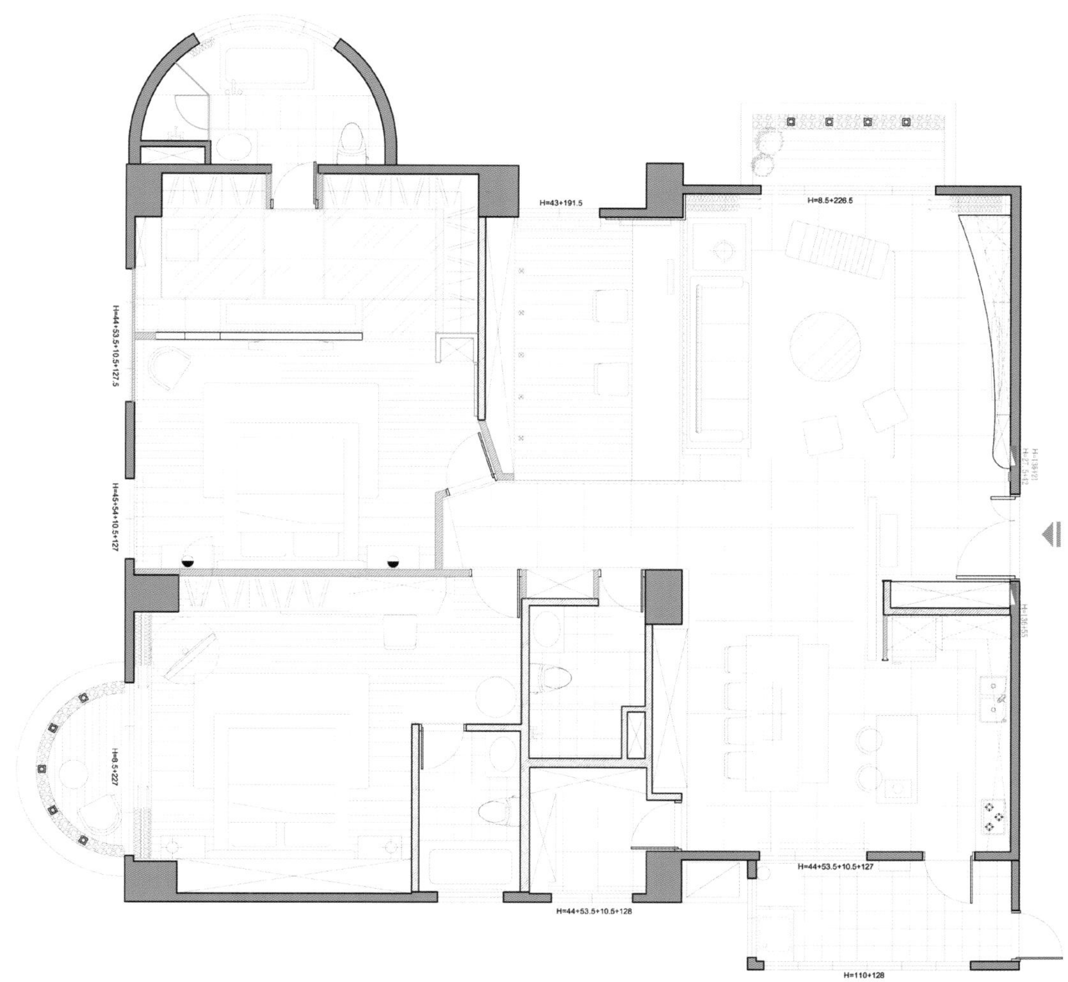

INFINITE CREATIVITY
创意无限想象

设计公司
怀特设计

设 计 师
林志隆

主要材料
清水板模漆、烤漆玻璃、黑网石、美耐板、墨镜玻璃、白铁板、鳄鱼皮革、马赛克砖、LED

摄影师
王基守

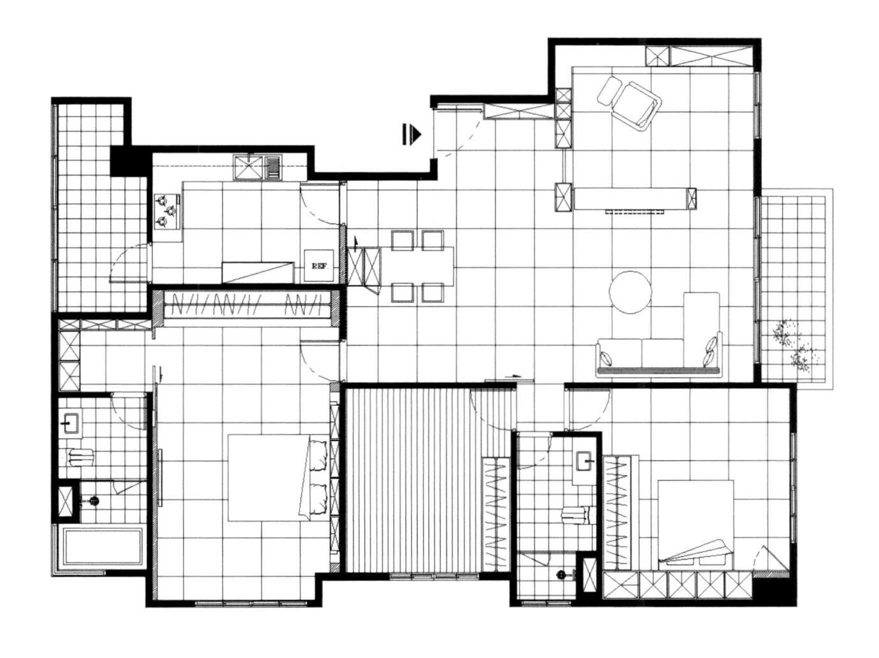

The door of the shoe box in an irregular shape echoes with the pictures on the facets of the ceiling, and the HL (Hairline) stainless steel which is used as the material of the door creates a sense of the technological future.

The television background wall is presented with large pieces of marble, but the ink painting grains in the marble reflect the green image of the space. Combined with the polished silica tile floor that is in light oyster grey, the background wall eases the abundant outdoor scenery, and makes the view more gentle and peaceful.

Looking for an imbalanced focus in the square space, the designer takes advantage of the bevel cut technique and makes it the focus of the whole design in order to break the cool lines and to create a new spatial configuration which forms an emotion beyond the cool feeling of buildings.

The black and white dining room, the paint glass wall, which is like the piano keyboard, extends straightly to the island table made of artificial stone and connects the two with each other. The table legs are made of HL stainless steel instead of tranditional materials. The layout, alternation and connection of modern materials promote the sense of fashion.

The master bedroom looks modest and steady, with simple lines and planes stretching from the living room in to create a simple and unique style of taste. The chocolate-block-shaped background wall of the bed with a cabinet hiding behind seems to be emitting sweet smell. Pointed rabbet joint piled layer by layer on the main wall creates a saturated sense in the three dimensional space, so that a plain surface is transformed into an artistic geometrical shape that spirals in from the outside to become the best decoration of the background wall. At the same time, the design of the background wall integrates a sense of balance well into the whole design and creates an aesthetic image.

鞋柜以不规则形的门片与天花板切割面的图形相呼应，门片材质运用毛丝面不锈钢切割营造出科技未来感。

电视墙饰以大片大理石，其水墨纹路映衬着空间内的灰阶影像，再搭配上淡染米灰纹的抛光石英砖地面，缓和了多样的户外景致，让视觉平和下来。

在方正中找寻到那不平衡的创意聚焦，斜角切割的手法成为设计焦点，打破了冷静的线条成为了空间的新造型，是冷白表情外的情感凝聚。

黑白主题的餐厅区域，宛如钢琴键盘的烤漆玻璃墙面，笔直延伸到人造石的中岛餐桌，将两种物件结为一体。餐桌以毛丝面不锈钢取代传统桌脚，现代化材质的铺陈、交替、衔接，加重了空间的前卫感。

以低调稳重为特点的主卧房，在形塑简约品位之际，线面创意也从客厅延伸至此。床头背墙呈现了巧克力砖的甜美意象，隐藏了后方的收纳柜体，借由层层相迭的勾缝企口雕刻主墙面在三度空间上的饱和性，单纯的平面转化为几何艺术形象，由外回旋入内化身为背墙的最佳妆点，同时也将整体设计的平衡感融入其中，整塑了美学意象脉络。

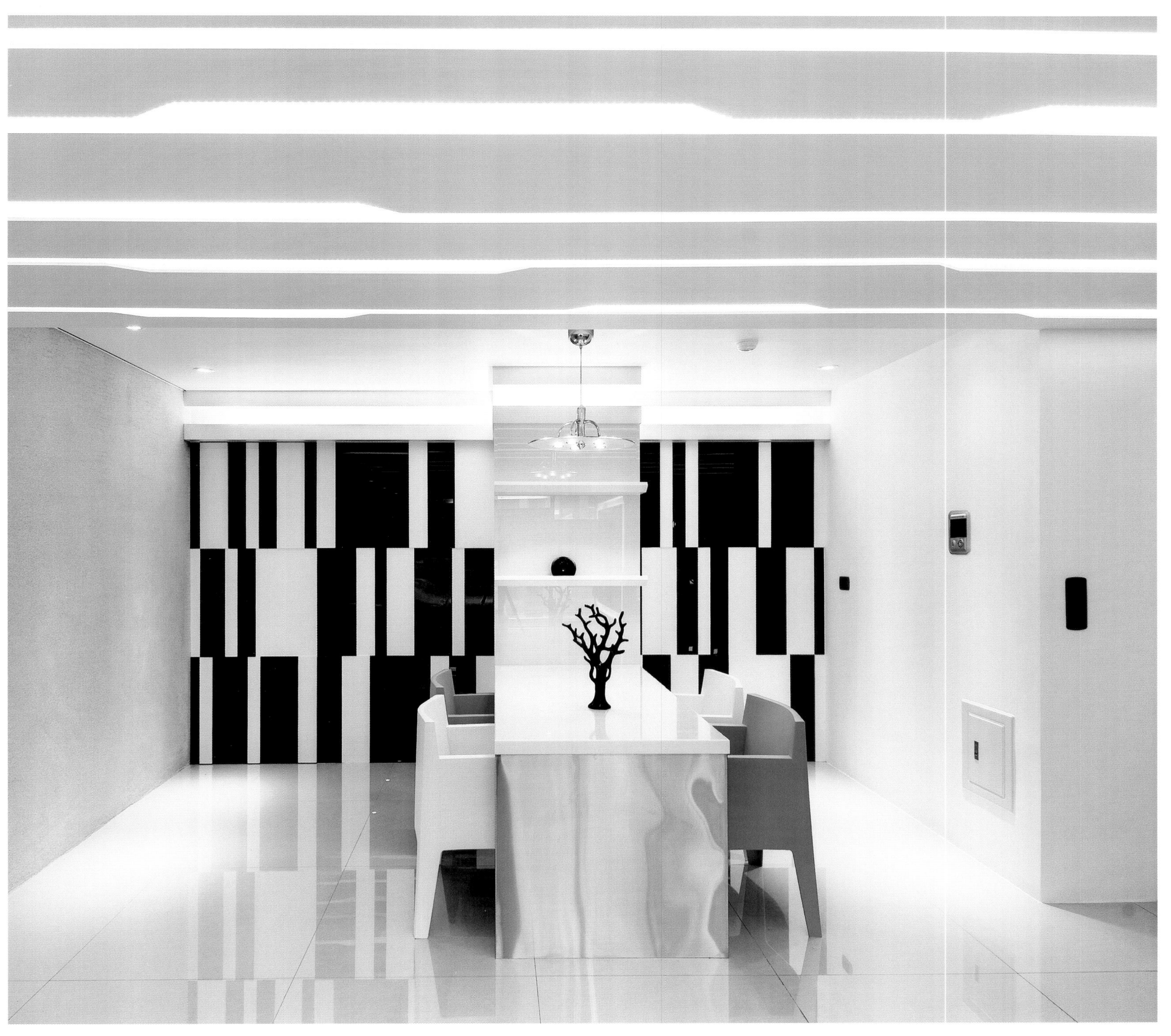

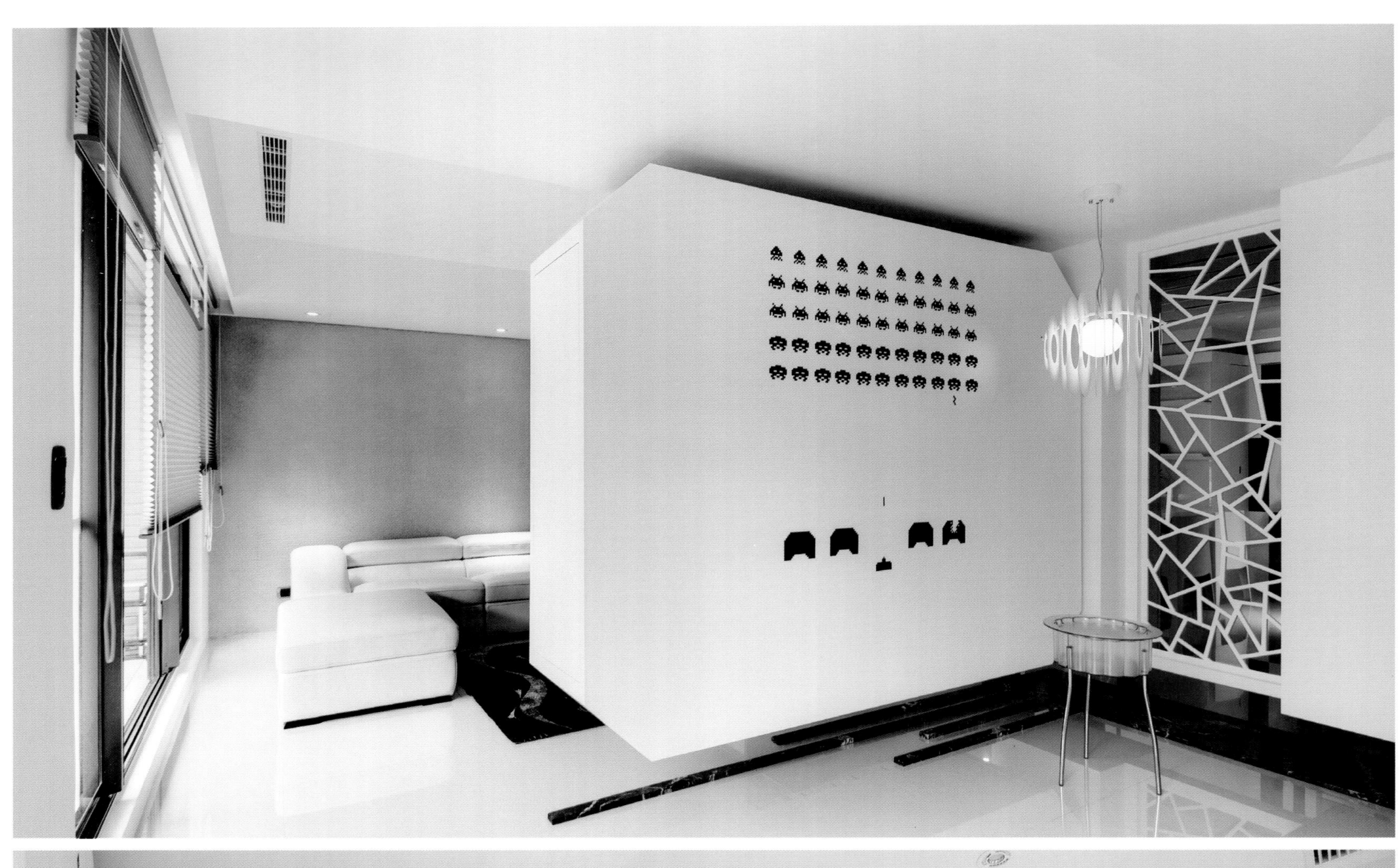

CONTENT HABITAT – MR WEI'S HOUSE

怡然居魏宅

设计公司
珥本室内装修设计工程有限公司

设 计 师
陈建佑、曾耀征

项目地点
中国台湾

项目面积
132m^2

主要材料
风化梧桐木、橡木地板、黑铁、镀钛板、乳胶漆、玻璃、3M 贴纸、石材

摄影师
吴启民

To let sunshine extend to every corner of the room, the designers have removed the partition wall blocking the sunlight and adopted light transmission materials to erase the sense of weight in the space. Functions of public space are superimposed, without obvious divided functional areas. The living room, dinning room and study equally take up three corners of the whole space to turn the precious space to good account, making indoor space less constricted and more capacious.

Natural lights penetrate through the transparent glass partition between the study and public bathroom to bring warm sunshine into every corner indoors during the day, striking a change of darkness from the color of weathering platane wood.

Apple green is set as the theme color of the space to echo with natural freshness. From the hallway to the TV wall and bedrooms, walls are also decorated with this color as an extension, in combinations with owner's favorite paintings of Jimmy, this kind of color assortment adds a sense of brisk relaxation and a little child interest into the space. Empty space in interior design is used to lessen the impact of visual change. Rough surface texture of weathering platane wood brings illumination changes, and soft wood texture also gives the best expression of natural and warm home style. Different materials of Ceramic tile pavement and oak floor permeate reciprocally to divide different functional areas. Black iron art items strike a sense of lines, meanwhile, the independent iron screen and its parts enrich light changes, attracting people involved in this three-dimensional space.

为了将光线延续到室内空间的每一个角落，我们拆掉了部分会阻碍光线的隔墙，改为采用透光性质的材料来轻化空间的重量感，并将公共空间的机能互相重迭，让彼此没有明显的分界，让客厅、餐厅、书房能融洽地分配着空间的三个等角，以此为空间争取更多的使用面积，也让室内不显得狭小。

自然光线透过书房与公共洗手间的透光玻璃隔屏，让室内的每个空间在白天都能感受到光线的温度。风化梧桐的木质颜色在光线的渲染下，也呈现出不同的深浅变化。

为了呼应清新的自然元素，以苹果绿作为空间的主题色。从玄关入口延伸至电视墙和卧室，搭配屋主喜爱的几米画作，通过高彩度的对比为空间增添了轻松的氛围与童趣。大量的留白缓冲了视觉变化上的冲击，风化梧桐的粗犷表面肌理让光线多了不同的层次，木质的温润也最能表达出自然与温暖的居家风格。地面上磁砖与橡木地板的交错延伸，区隔出不同的机能空间。黑色铁件增强了空间的线条感，独立的铁件屏风与构件，还能丰富光线的阴影变化，强调出立体感。

WINDOW DESIGN
SHOP DESIGN
WOOD DESIGN
DOOR DESIGN

YUANLI CENTURY CONVERGENCE CENTER, A3-8F

元利世纪汇 A3-8F

设计公司
异国设计

设 计 师
王文亚

项目地点
中国台湾

项目面积
130m²

The project is located in the center of Taipei City, a residential building with its focus on cultural and educational atmosphere. Its design blueprint is based on the theme of the humanity and sets as a model of fashion style, blending its humanity feature into excellent spatial framework to create a more capacious level arrangement. Different from the common design style which is with classic and luxurious tone, this design takes on a sense of fashionable and exquisite taste and graceful nature. The mixture of visual appeal and its spatial framework constitutes a comfortable living space, affording an atmosphere which is full of abundant sunlight and comfortable capaciousness.

When you step into the porch, what comes into your view first is the elegant porch ground. A variety of different building materials shape up a clear and capacious vision to correspond with what should be an attitude and value in this cultural and educational space. The using of wood skin-color is extended into the living room, dinning room and bedrooms. Dimension radiating from point, line and plane creates a design and constructs several independent spaces as well as interdependent ones. The designer particularly chooses furniture with conciseness rather than luxury to create a warm and succinct housing aesthetics. Steady materials with light or dark coffee colors add a sense of relaxation and elegance into the whole residence. The use of proper proportion, dimension and materials not only combines space with its practical functions but also brings a visual impact and adds a cordial feeing.

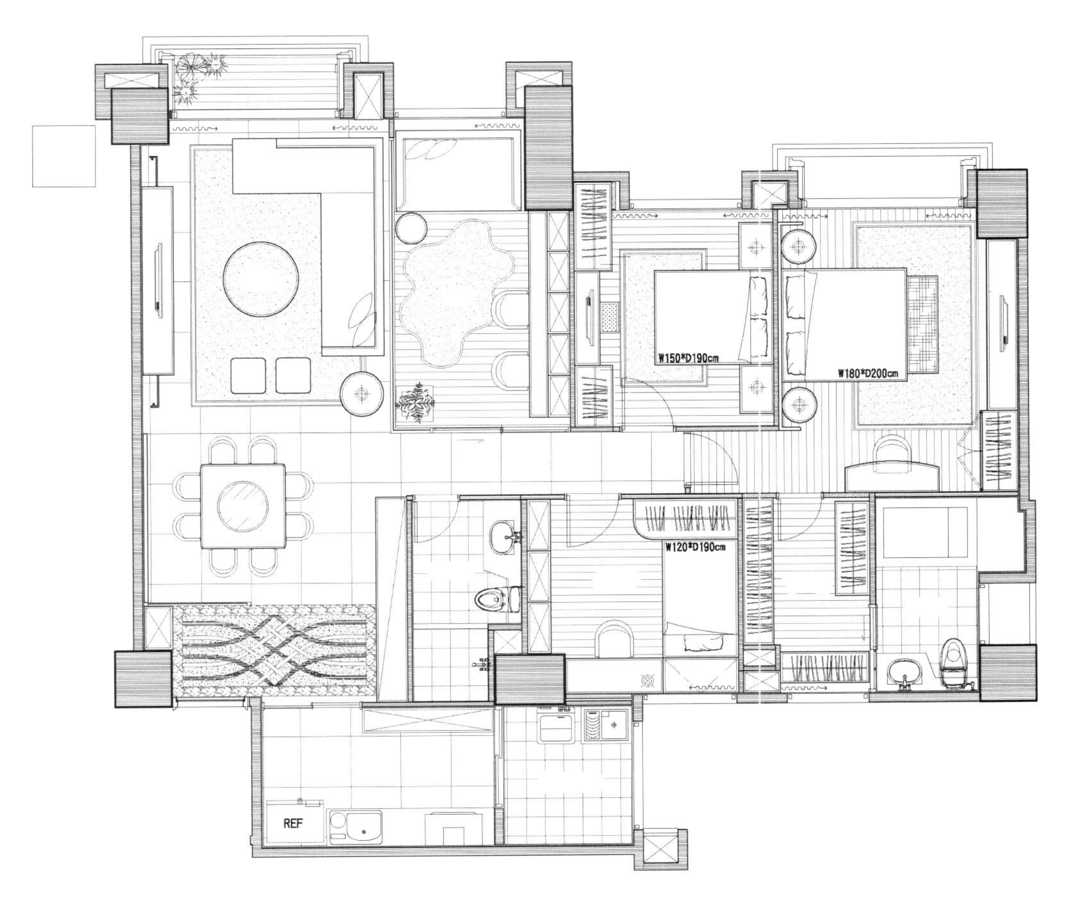

本案位于台北市中心，一座以人文价值为主轴的小区内，此案的设计蓝图以人文主题为发想，并赋予其时尚的风格典范，将精品架构与人文气质相结合，营造出更多的空间层次。不同于一般设计风格中的古典华丽，此整体设计呈现时尚及质感的品位与雍容，融合后的视觉语汇与空间架构成为舒适的生活空间，并且提供了充沛的光照与舒适的空间氛围。

甫进玄关，映入眼帘的是典雅的玄关地面，多元的建材构造出利落的宽敞视野，呼应了人文空间应有的态度与价值。设计将木皮色调延伸至客厅、餐厅和卧室，由点、线、面三个方向拉出呼应的设计感，构建了数个既独立又相互依赖的空间。设计特别选用简约大方的质感家具，不是炫耀奢华，而是为了营造出温馨大方的新住宅美学。深浅咖啡及沉稳的质材成为空间轻松优雅的催化剂，比例、尺寸及素材的运用，将空间及其实用机能整合起来，提升出视觉冲击力，又体现了亲切感。

ABSOLUTE BEAUTY
纯粹美好

设计公司
宽月空间创意

设 计 师
吴奉文、戴绮芬

项目地点
中国台湾

项目面积
137m²

主要材料
盘多磨、檀木、橡木、杉木、印度黑、铁件、黑玻、德国环保涂料及接合剂

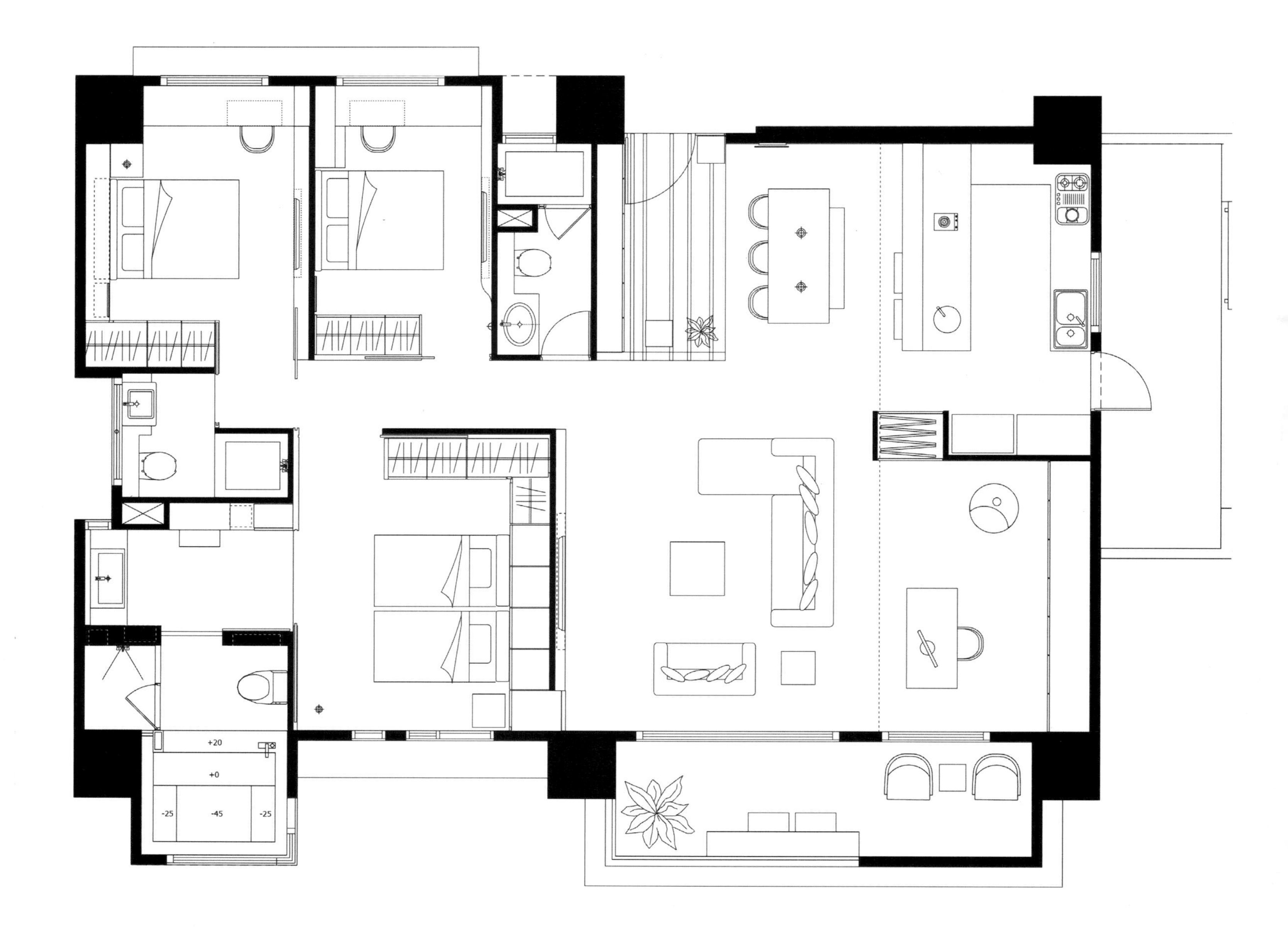

The project tries to convey a simple and stable atmosphere by the use of colors that are not in sharp contrast, like warm dye grey, beige and silent black. Meanwhile, the concept of "less decoration, more ornament" is stressed so that modeling elements and lines are seldom used to create a space with cultural atmosphere rather than the one that runs after trend. Therefore, the materials of the furniture and decorations are properly processed to echo with the inhabitant's traits.

The L-shaped leather couch, the handmade rattan double chair and the tea table in peculiar geometrical shape in the living room, combined with the classical single armchair and the rattan-weaved hanging seat in the study, create a modern and relaxing atmosphere. British brand floor lamp and desk lamp are in the color of foggy gold, which conceals part of the splendid luxury. The wood floor of the outdoor terrace has been dyed into grey – a color of antiquity, which makes the floor look natural as if it has experienced the passage of time. The unique outdoor bar is actually a decoration to conceal the survival equipment on the terrace because the project is on the emergency floor and the survival equipment cannot be discarded. The design of the bar on the one side ensures the function of the survival equipment, while on the other side it provides the owner with a relaxing area of afternoon tea or barbecue. There are too many possibilities in our life, but the essentials of life should always be beautiful.

全案没有强烈的对比色彩，温暖的染灰木色、米色配上沉稳的黑色，传达出简练稳重的氛围。同时，设计运用轻装修重装饰的概念，删减过多的造型、线条，让空间不追随潮流而深具人文气息。因此，设计对家具、家饰的材质、色彩做了相应的处理，以呼应居住者的特质。

客厅 L 形沙发具有皮革质感，搭配手工藤编双人座椅，茶几投射出特殊的几何形图腾光影，书房里选用具有扶手的古典单椅，结合藤编吊椅，塑造出既现代又悠闲的气氛。英国品牌雾金色的立灯和桌灯，隐藏着些许华丽的质感。户外阳台部分，木地板经过染灰处理，带着仿旧质感，呈现出历经岁月洗礼的自然感；而独具风味的户外吧台的存在，其实是因为该住宅属于逃生楼层，无法舍弃阳台逃生设备，因此，在不影响逃生设备的功能下，设计了吧台，这一举措也为屋主创造出午茶小憩、烤肉的小空间。生活中有太多可能，而本质就该是如此纯粹美好。

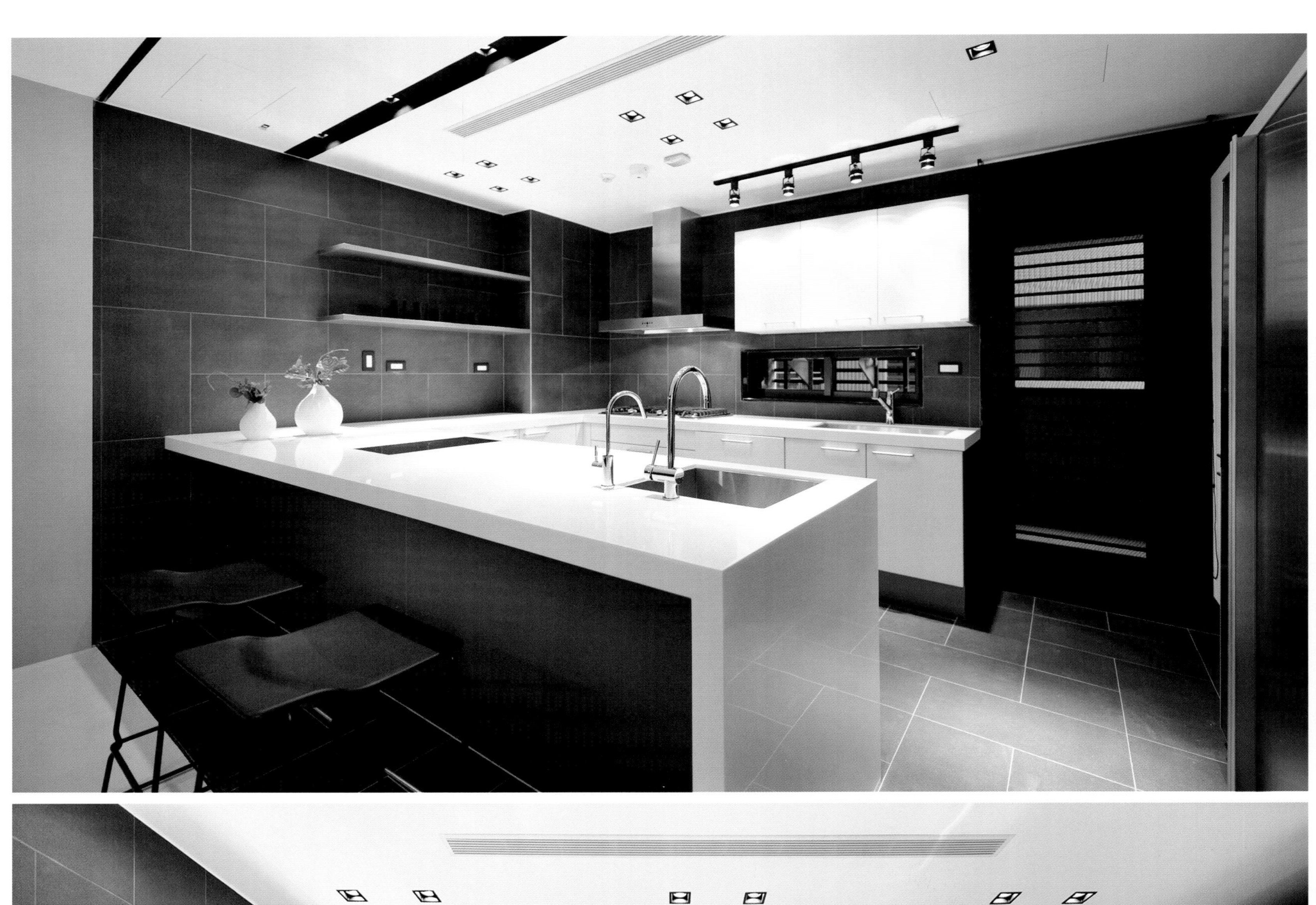

DOWELL ESTATE – XIANGSHAN A2 SHOW FLAT

东原地产——香山 A2 样板房

设计公司
德坚设计

设 计 师
陈德坚

摄影师
陈维中

The concept of this project is to create a brand new feeling based on a unique modern and concise style, in order to present refined appearance through simple design and to illustrate fashionable taste, making space comfortable and modern. Though the show flat was constructed in a concise style, the design of every detail is with dedication and carefulness. The designer uses simple interior lines to create a concise, bright and clean profile without complicated structure or arrangement.

This concept has been applied to every part of the space, including the kitchen, the washing room and the living room, displaying an ingeniously designed residence. The modern residence style for common people is thus presented without extra or contrived decoration, and the climax of aesthetics is thus achieved.

Life tastes are conveyed through living space, as well as objects endowed with meaningful affection decorated in the space(such as several potted green plants), which definitely offers an oasis for relaxation after a day of hard work. In the residence, proper human-orientated small designs scattered in the space, which creates a comfortable and harmonious environment in every living area.

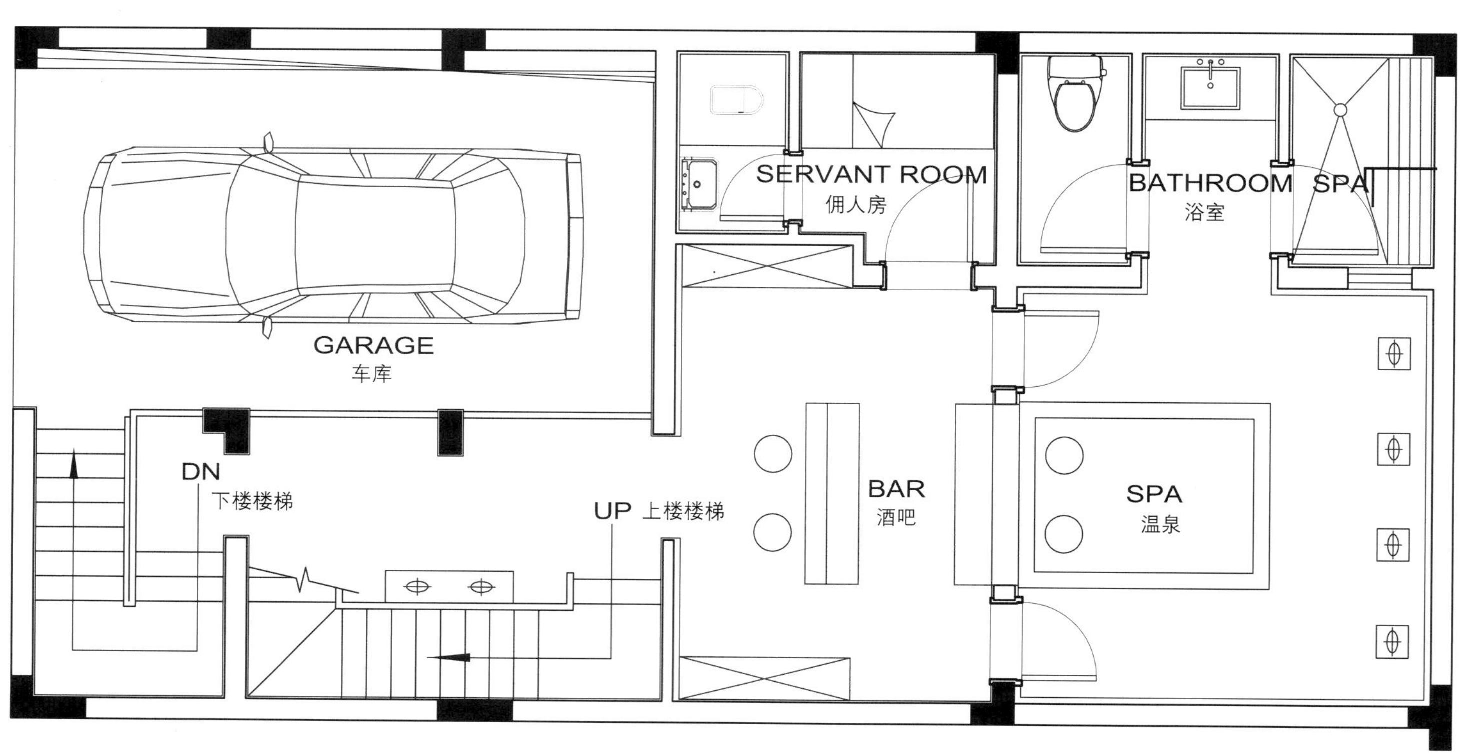

G/F LAYOUT PLAN
底层平面图

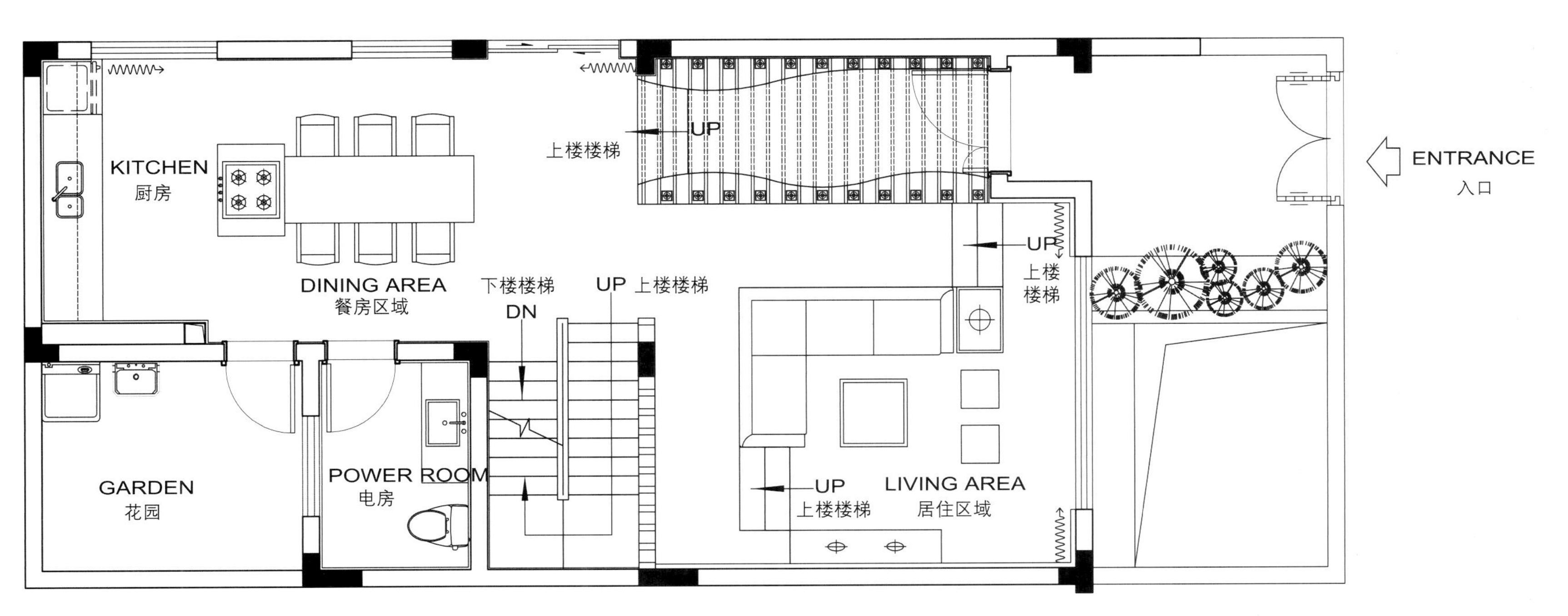

1/F LAYOUT PLAN
一层平面图

本案的设计理念是以独特的简约现代风格，为不同的空间赋予全新的感觉。在简单中呈现细腻，给人一种温馨舒适的现代感，并凸显时尚品位。虽然样板房以简约风格为主，但设计处理每一处时都非常细致用心。设计师尽量简化室内的线条，没有繁复的结构和布置，从而令空间简洁而明净。

不论是厨房、卫生间，还是客厅，设计师都营造了这种感觉，让整个房间变得轻巧。没有额外或刻意的修饰，却能展现一般现代人的家居风格，达到美学的最高境界。

将生活品位表达于家居中，将一些具有情感意义的对象布置于空间中（如几盆翠绿的盆栽），让人在紧迫的城市生活节奏下享受那难得的闲暇。于设计上再加上对人的体贴，点到即止，就能为生活的每个角落营造出舒适与和谐。

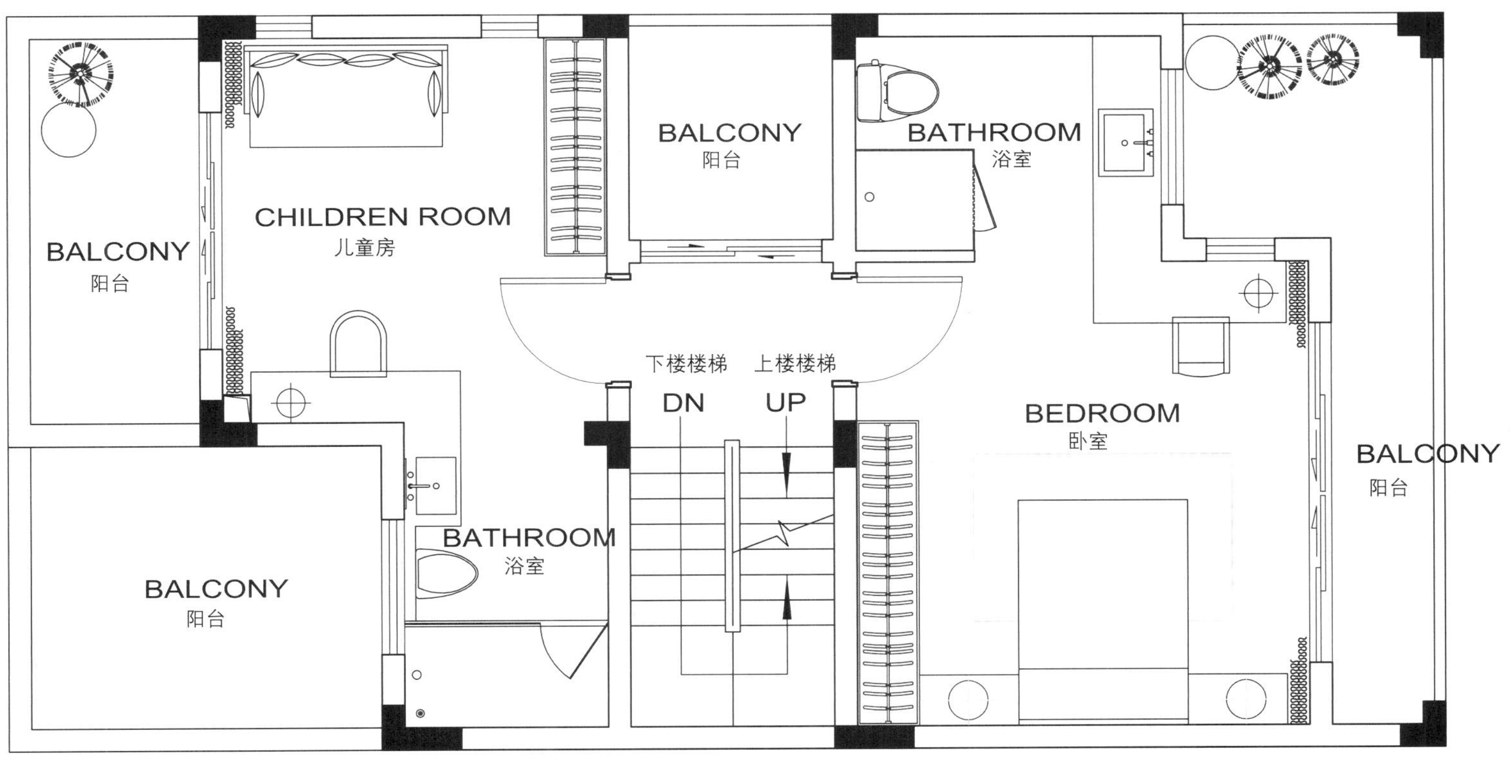

2/F LAYOUT PLAN
二层平面图

SPACE
Homidea

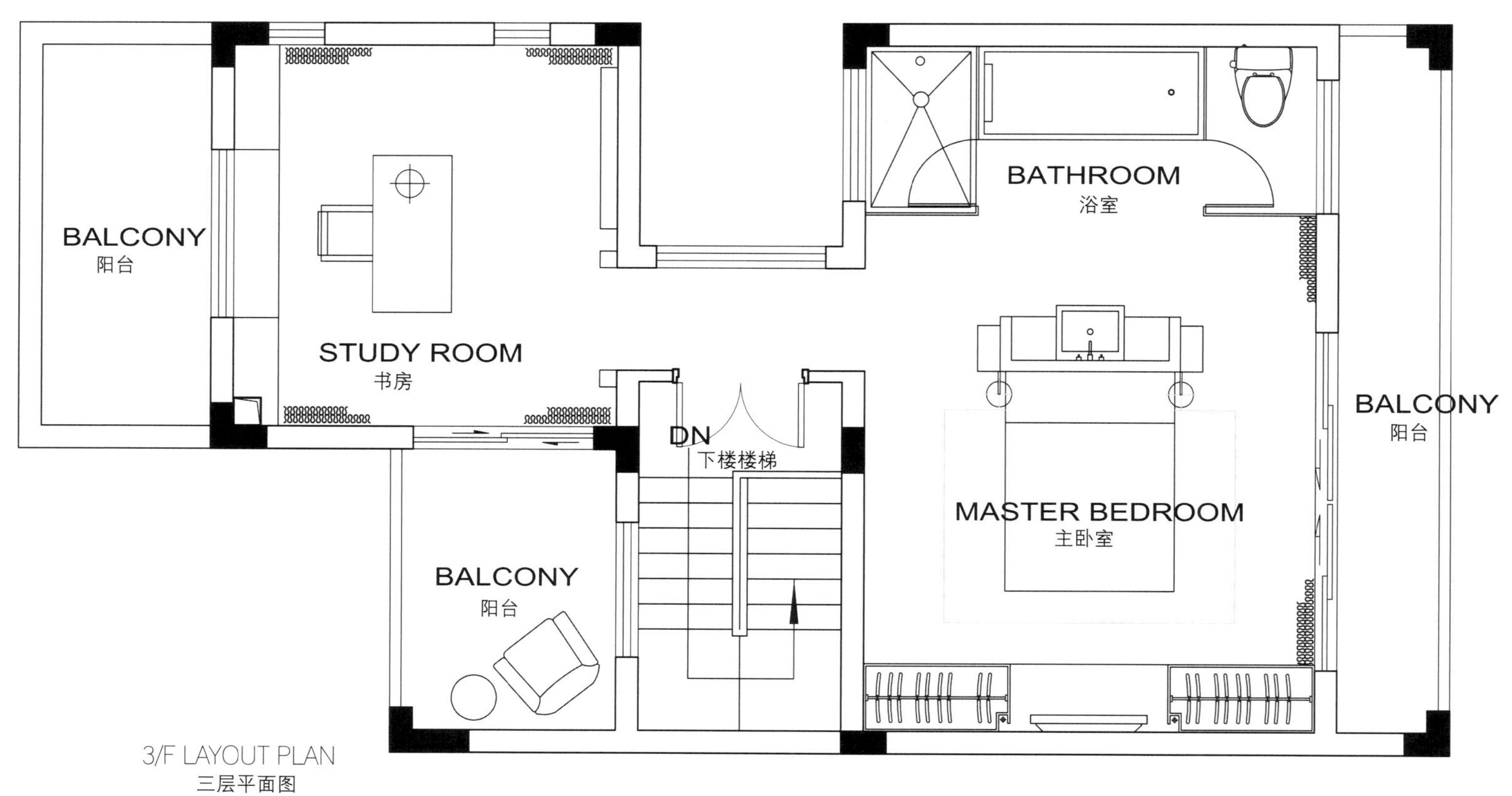

3/F LAYOUT PLAN
三层平面图

THE PENTHOUSE AT 17 MILES

十七英里

设计公司
德坚设计

设 计 师
陈德坚

项目面积
250m²

摄影师
陈维忠

In the 3-storey residence, the suspended wood stair-way extends from the marble wall of the basement to the first floor matching up perfectly with the transparent railings made of tempered glass, and forms a set of overlapping geometrical lines.

The suite on the first floor has a terrace outside, and inside the room, the bed back is a piece of clear mirror which reflects the whole space while the end of the bed is a partition wall made of tempered glass which breaks through the tradition of using mirror in the bedroom. Meanwhile, the tempered glass also serves as a connection of the view of the first floor and the basement. The washroom is also designed in an unusual pattern in which the toilet and the shower are separated so that the function of the bathroom is divided. The hand basin therefore is moved near the room, combined with the double-sided mirror to create a unique washing area.

The master's suite in the third floor is connected directly to the open study. In the room, the leather bed and leather furniture go well with the brown tile and bed backplane. The skylight on top of the desk is designed so delicately that the starry night sky can be seen clearly through it, providing the owner with a place to appreciate the beauty of the nature, to relax entirely and to stay away from all the trouble. The dressing room on the other side of the suite decorated with gray marble is another highlight of the space. Integrative bathtub and wash basin match well with the wall and floor which use the same material offering a simple look. Staying in the bathroom, one cannot help longing to have a warm bath, to walk out to the terrace barefoot, and to stand on the marble floor with small holes to have a water massage on the soles.

此案由三个楼层组成，地下至一楼的悬浮式木梯从云石墙中伸展出来，配合通透的强化玻璃围栏，形成一组重叠出现的几何线条。

设于一楼的套房连接露台，床背板以清镜反映空间，床尾则以强化玻璃作为间墙物料，既打破了将镜子用于睡房的传统，又可衔接一楼与地下楼层的视野。套厕格局亦不常见，独立分隔的坐厕区及淋浴区，既将浴室功能一分为二，又将面盆移近房间，配合双面镜子设计，营造出不一样的卫浴环境。

三楼的主人套房连着开放式的书房，以皮制睡床及家具搭配啡调的墙砖床背。书桌顶的天窗设计最为特别，以天窗引入闪闪发光的星空，能让屋主在假期中静下来欣赏自然环境，抛开身心烦恼。与主人房左右分隔的套厕连着衣帽间，以卡撒灰云石营造出豪宅的另一焦点。一体化的面盆连浴缸，配合同一物料的墙身及地台设计，视觉上非常简洁。置身于浴室内，无不想享受一下窝心的热水浴，再赤脚走出露台，踩在布满小孔的云石地台上，让流出来的水按摩脚掌。

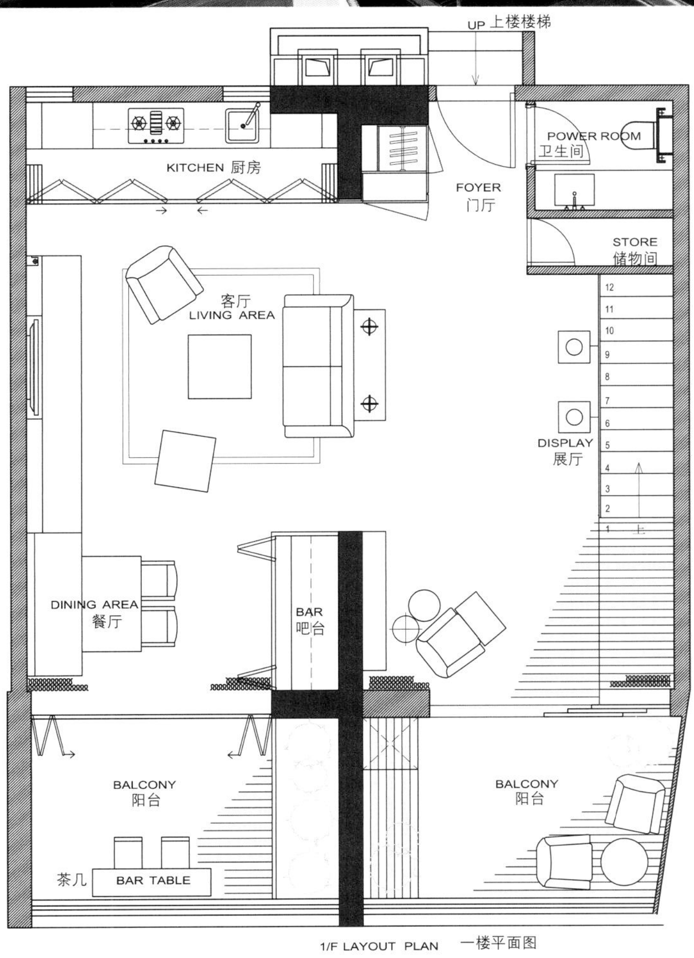

1/F LAYOUT PLAN 一楼平面图

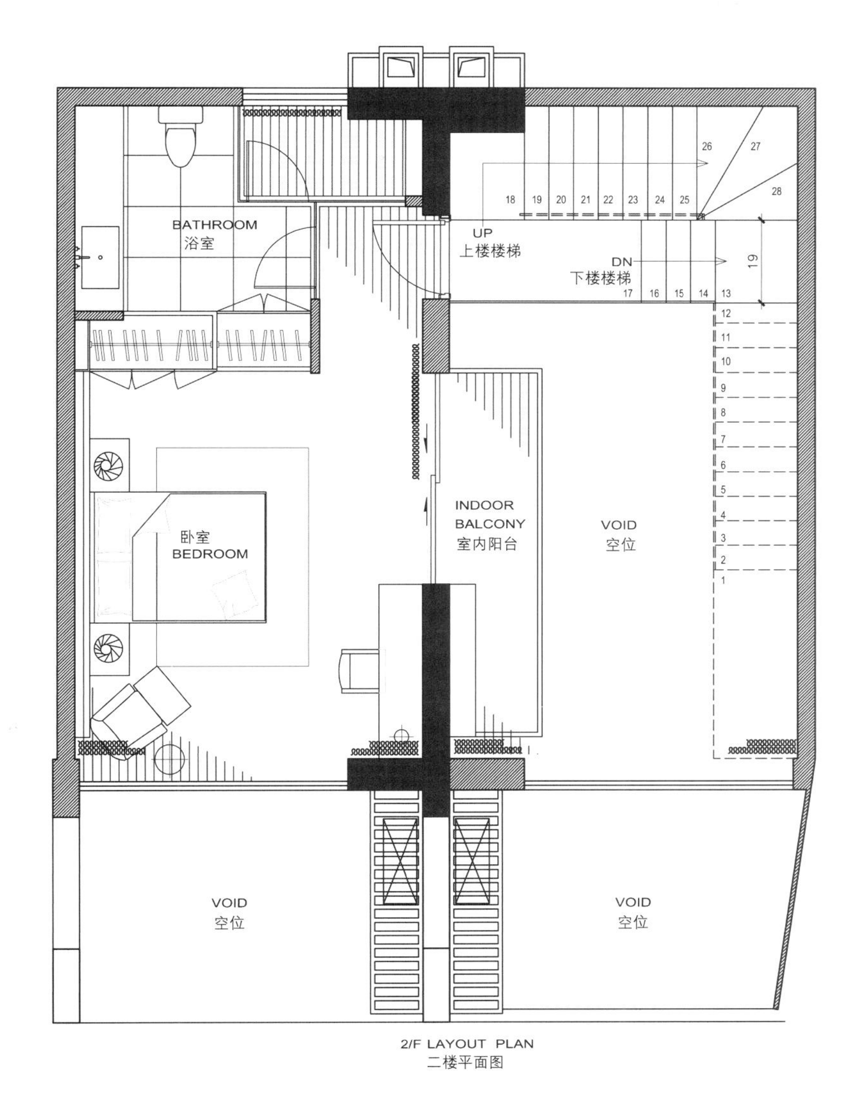

2/F LAYOUT PLAN
二楼平面图

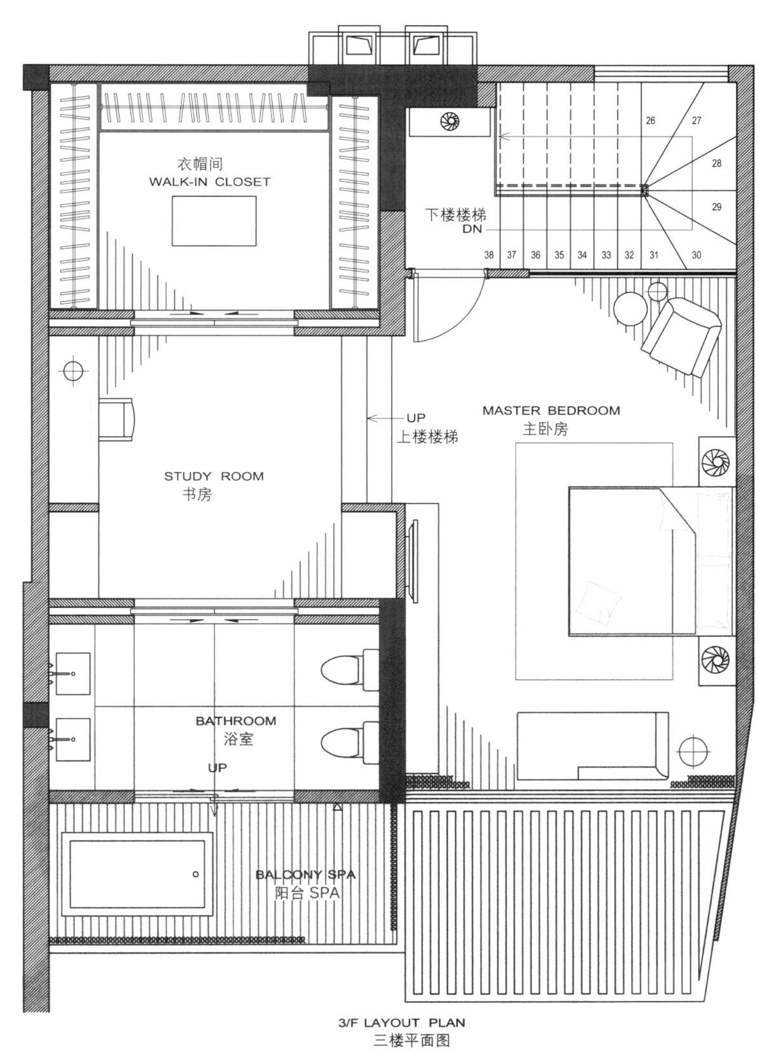

3/F LAYOUT PLAN
三楼平面图

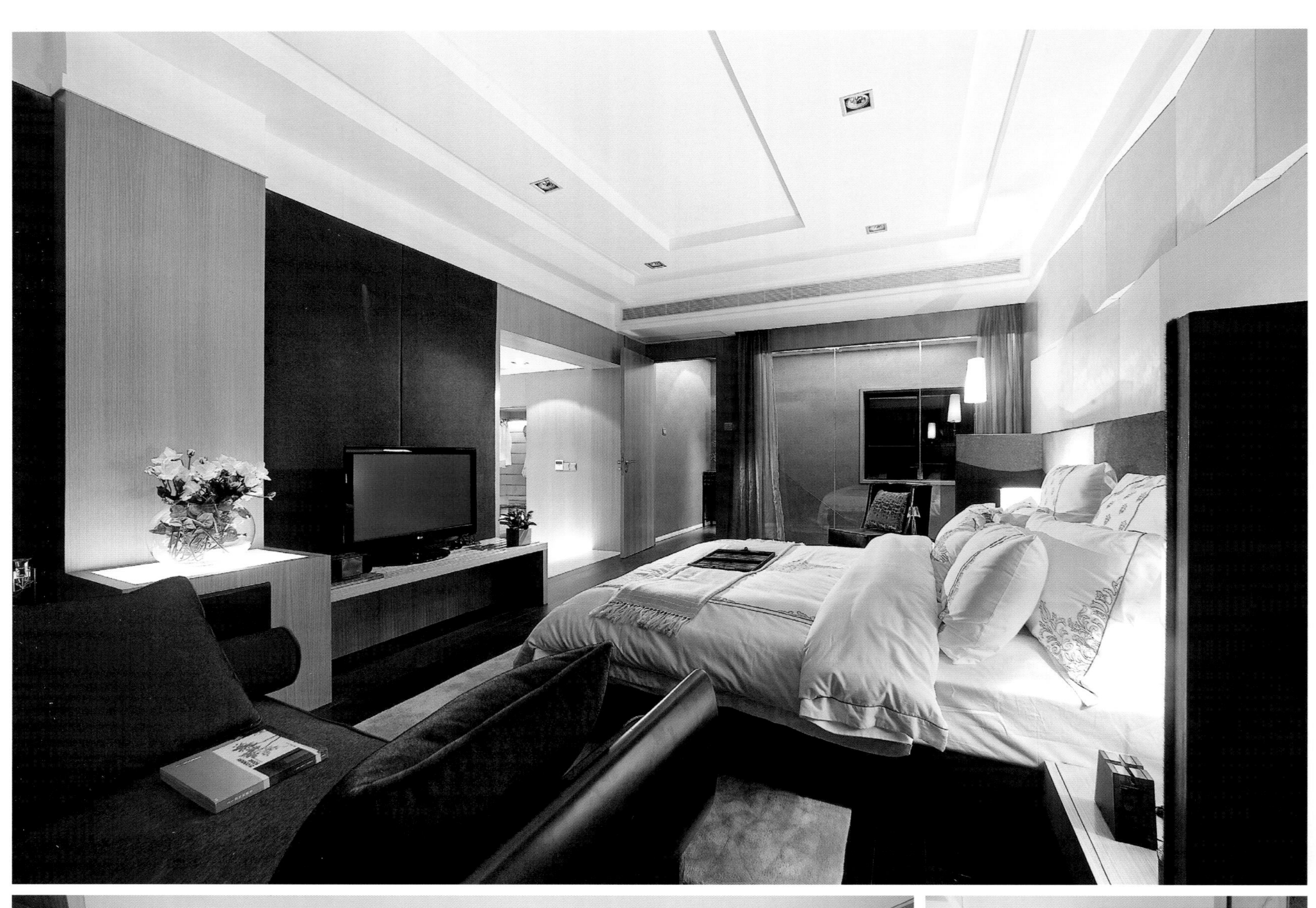

MR ZHENG'S RESIDENCE IN TAIPEI

台北郑宅

设计公司
玮奕国际设计工程有限公司

设 计 师
方信原、洪于茹

项目地点
中国台湾

项目面积
240m²

主要材料
砂岩大理石、柚木、海岛型复古刻沟木地板、不锈钢、强化茶色玻璃、黑色烤漆玻璃、进口壁纸、橡木实木、冷喷漆、钢琴烤漆

Once stepping into the house, you will notice that on the left side of the room, a tall white cabinet with an extremely simple appearance is placed against the wall, matching with the suspended iron partition screen. Arrayed grille above the channel enhances the visual effect. Besides the lights, the grille itself is in a banked angle to create a visual effect of deflective parallels. Opposite to the entrance, there is a display counter designed for the collections with a background wall in iron grey. The facades are connected with an oblique technique. The rear part is a store room which on the one side saves several cabinets and on the other side makes a good preparation for the changing space.

Behind the couch of the living room, there is a wood side cabinet, which is separated from the tall white cabinet against the rear wall. Under the tall cabinet, grey marble is designed into different layers, where the collections can be displayed. The television background wall in a sculpture effect is the visual focus with oblique lines creating a simple atmosphere. Color blocks in iron grey extends from the walls of the terrace to the display wall at the end of the corridor, the step by step design in different width makes the blocks look like deformed three-dimensional paper folding works.

The walls in the corridors leading from the public area to different rooms are painted with delicate technique which transforms the simple planar lines into three-dimensional effect. The multi-function room looks delicate with the design of folding door, floor levels and diverse application of collocated materials.

The main bedroom is at the end of the corridor. The comfortable sense is enhanced by American styled fresh colors, white color, classical lines and shuttered window. The door of the inside bathroom is decorated with classical lines in dark colors which helps stabilize the space. The wood floor and separated bathtub inside soften the toughness of the grey color and enrich the visual texture of the space.

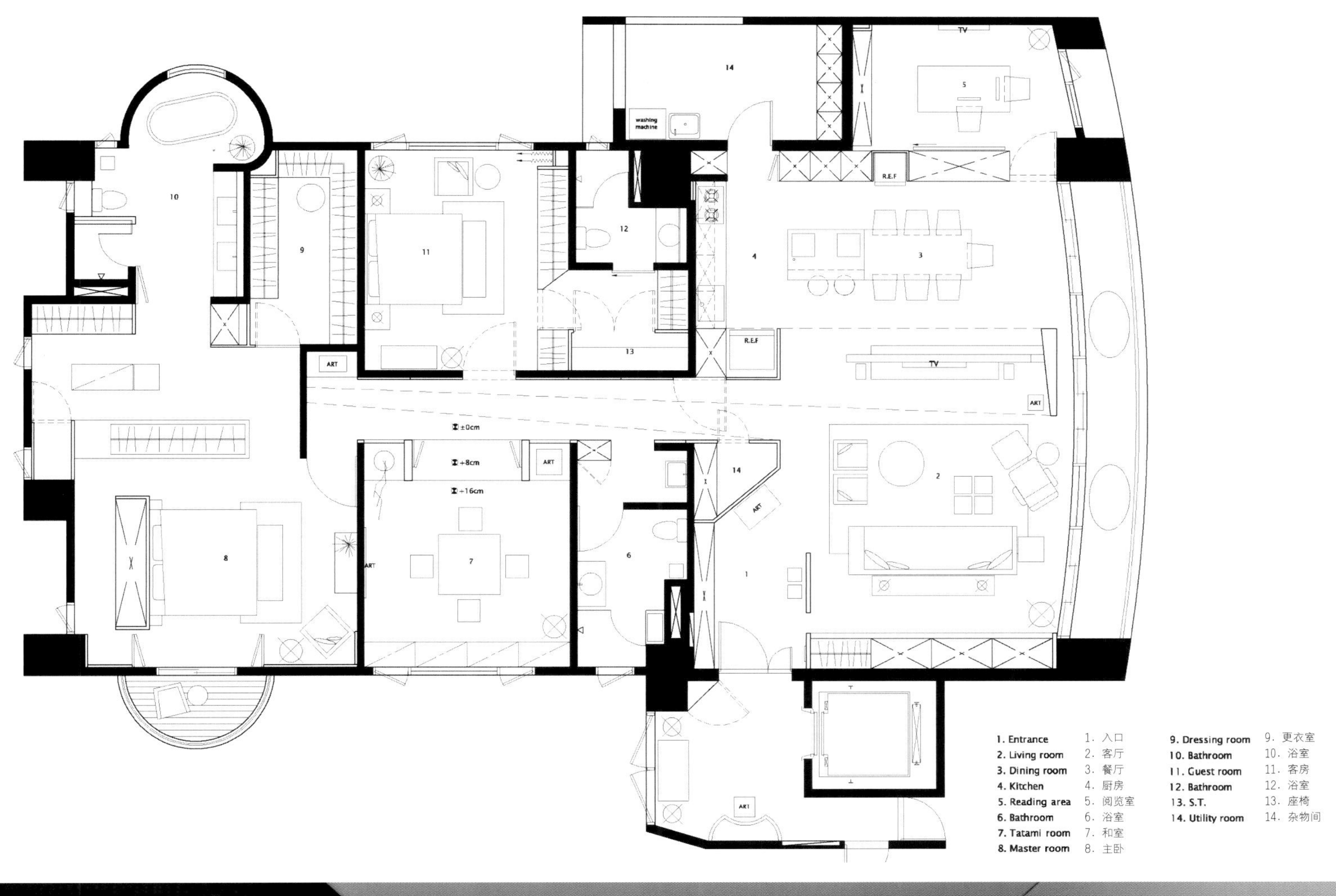
1. Entrance 1. 入口
2. Living room 2. 客厅
3. Dining room 3. 餐厅
4. Kitchen 4. 厨房
5. Reading area 5. 阅览室
6. Bathroom 6. 浴室
7. Tatami room 7. 和室
8. Master room 8. 主卧
9. Dressing room 9. 更衣室
10. Bathroom 10. 浴室
11. Guest room 11. 客房
12. Bathroom 12. 浴室
13. S.T. 13. 座椅
14. Utility room 14. 杂物间

进入室内，左侧靠墙处为外型极简的白色收纳高柜，衬托出悬吊结构的铁件隔屏。通道上方以行列式格栅增加视觉效果，除灯光搭配外，格栅本身还有角度上的偏斜，营造出线条平行间的落差变化。正对入口处有为收藏品制作的陈列台，背景墙为铁灰色调，采用偏斜手法作为立面的衔接，后方为收纳室，既避免了空间中出现过多的实体橱柜，又为空间转折处预留了伏笔。

客厅长沙发后有特制的实木边柜，与后方靠墙的简约白色高柜相搭配，高柜下方运用灰色砂岩大理石，塑造出低调渐层效果，同时还可陈列屋主的收藏品。极具雕塑感的电视主墙是视觉的焦点，以斜线切割的轻薄量体搭配极简的氛围。铁灰色调的色块从阳台处墙面开始延伸至长廊底端的展示墙，分段不等宽的设计以一种变形折纸般的立体概念运作开来。

公共空间通往其他房间的过道处，以精湛的立面勾缝技巧，让简洁的平面线条呈现出立体感。多功能房以折门、地板落差及材质运用搭配上的变化营造出精致的空间感。

主卧房位于长廊尽头，房内以清爽色系、白色系、古典线条、百叶长窗等美式风格底色来提升舒适度。主卧浴室入口门上点缀些古典线条，以深色调来稳定空间，内搭实木地板、独立浴缸，柔和了灰色调的刚硬，丰富了空间的质感层次。

GUOTAI TIANMU SHOW FLAT

国泰天母样品屋

设计公司
动象国际室内装修有限公司

设 计 师
谭精忠

参与设计
詹惠兰、陈敏媛、何芸妮、黄利画

项目地点
中国台湾

项目面积
510m²

主要材料
喷漆、镀钛、钢刷木皮、水晶马赛克、壁布、钢刷木地板、石材、皮革夹纱玻璃、灰镜、墨镜、雪花石

摄影师
庄孟翰

Entering the living room and the dining area through the hallway, one will be impressed by the spacious and decent feeling filled in the open area of the living room and the dining area. 10-meter-wide space view is an excellent representative of a mansion. The main wall of the living room is the combination of travertine and an inbuilt fireplace with simple lines. The wall surface with the texture of steel brush and veneer, and the skirting and the ceiling margin decorated with crystal mosaic and titanium coating create a unique visual impression.

Open ended dining hall combines perfectly with axis of the central working area of the light meal kitchen, making the space consistently smooth. The design of ceiling lighting is realized with the material of gauze glass which permits gentle light out. Complete cooking appliances and kitchenware make both light meal and festival get-together possible, letting guests enjoy a delightful meal time.

The indoor hallway creates a completely independent and integrated place for artistic appreciation.

The master's room still stresses fashion and comfort. The display cupboard made of steel brush and veneer, matched with the crystal mosaic and titanium coating, makes the bedroom delicate and fashionable.

The dressing room can store articles like trunks and quilts. The flannelette grids in the drawer are thoughtfully designed so that the owners can put in their personal valuables like jewels, watches, and glasses etc.

In the main bathroom, the equipments are highlighted by the black mesh stone. In addition, titanized mirror frame and the alabaster wall lamp make the overall atmosphere noble and mysterious.

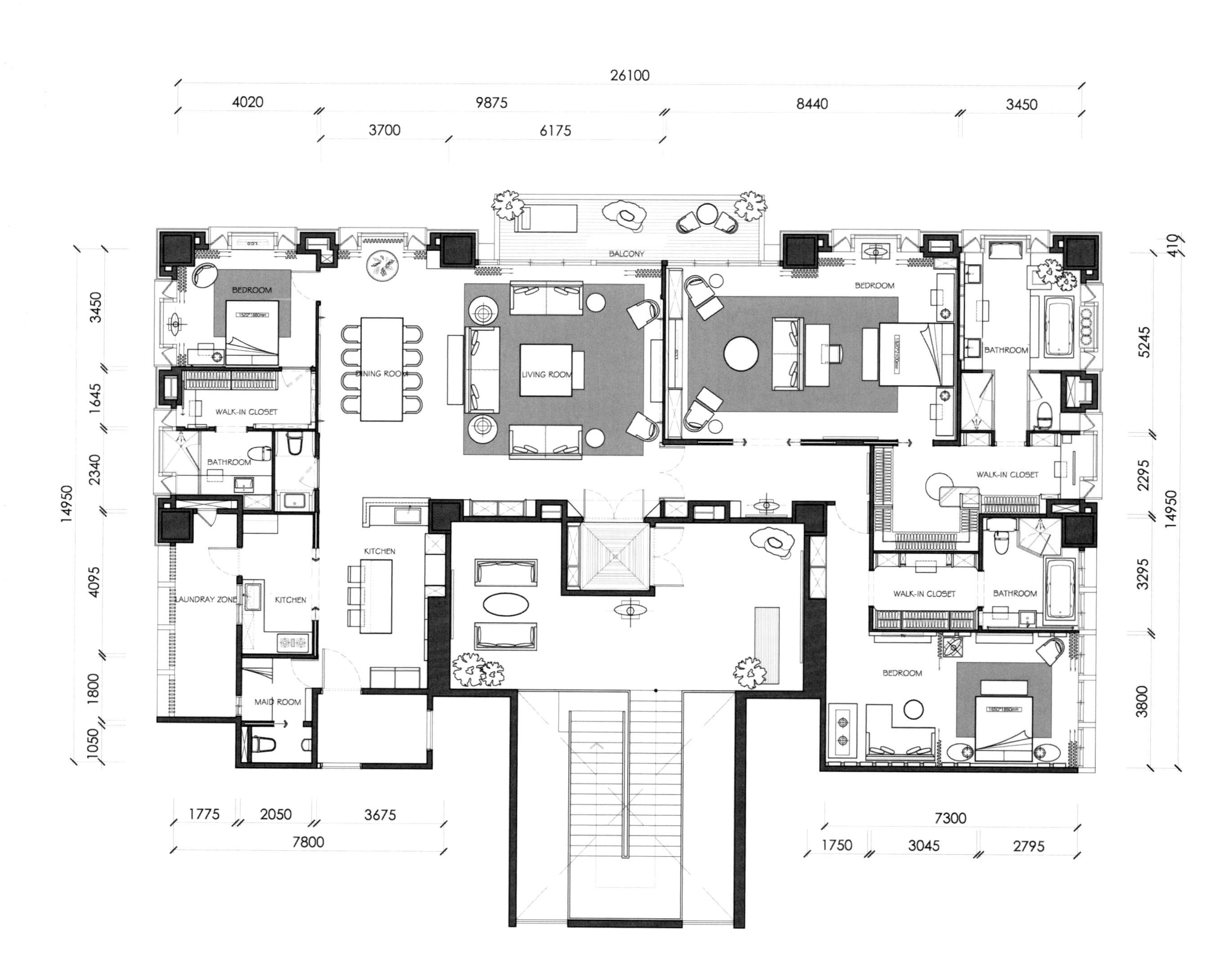

由外玄关进入客厅、餐厅时，立刻就能感受到贯穿客厅与餐厅的开放式空间的宽敞与气度，约 10 米宽的开阔的空间视野，呈现出大器豪宅的气势。客厅主墙以洞石搭配线条简单嵌入型的壁炉，钢刷木皮质感的壁面造型，踢脚与天花收边在水晶马赛克与镀钛的点缀下，创造出独有的视觉韵味。

开放式的餐厅搭配了轻食厨房中岛的轴线设计，除空间一贯的流畅外，天花板的照明设计更以夹纱玻璃材质规划，渗漏出柔和的灯光，搭配齐全的料理设备，不论是轻食还是大宴，都能使宾客恰如其分地享受美食风味。

内玄关于住宅内创造出一个完全独立、完整的艺术观赏空间。

主卧室的设计维持了一贯时尚、舒适的基调，钢刷木皮的展示柜及天花板，搭配水晶马赛克与镀钛的点缀，再次呈现出主卧室的精致与时尚。

更衣室可收纳行李箱、棉被等物品，抽屉内贴心地设计了绒布格，以便男女主人放置首饰、手表、眼镜等贵重物品。

主浴室以黑网石石材衬托高质感的卫浴设备，镀钛的镜框搭配雪花石壁灯使整体氛围更显高贵神秘。

SAMPLE RESIDENCE IN ZHUBEI

竹北样品屋

设计公司
杨焕生设计师事务所

设 计 师
杨焕生、郭士豪

项目地点
中国台湾

The project is a private club on the open river bank with a superb natural environment which offers full and extensive horizon of the spatial river scenery on one side. The lighting design is based on the river view with large windows that permit a direct communication between the large space and the nature. The concept of the design is based on "watching", which makes the visual effect of the indoor and outdoor space similar to that of an artistic gallery.

The space spreads from a corridor to both sides, and the reserved allocation highlights the spacious space above and the zones below. Grey mirror is used as the material of the wall to present a natural texture like water waves, creating an area in which dramatic effects and sensory experience coexist.

The corridor is the central axis of the space, which compresses the height and width of the space into a walking scale. Accompanied by the works of art on both sides, the axis is gradually and peacefully penetrating into other spaces, the design then brings the river and the valley back in front of the owner through the penetrating space.

CENTURY

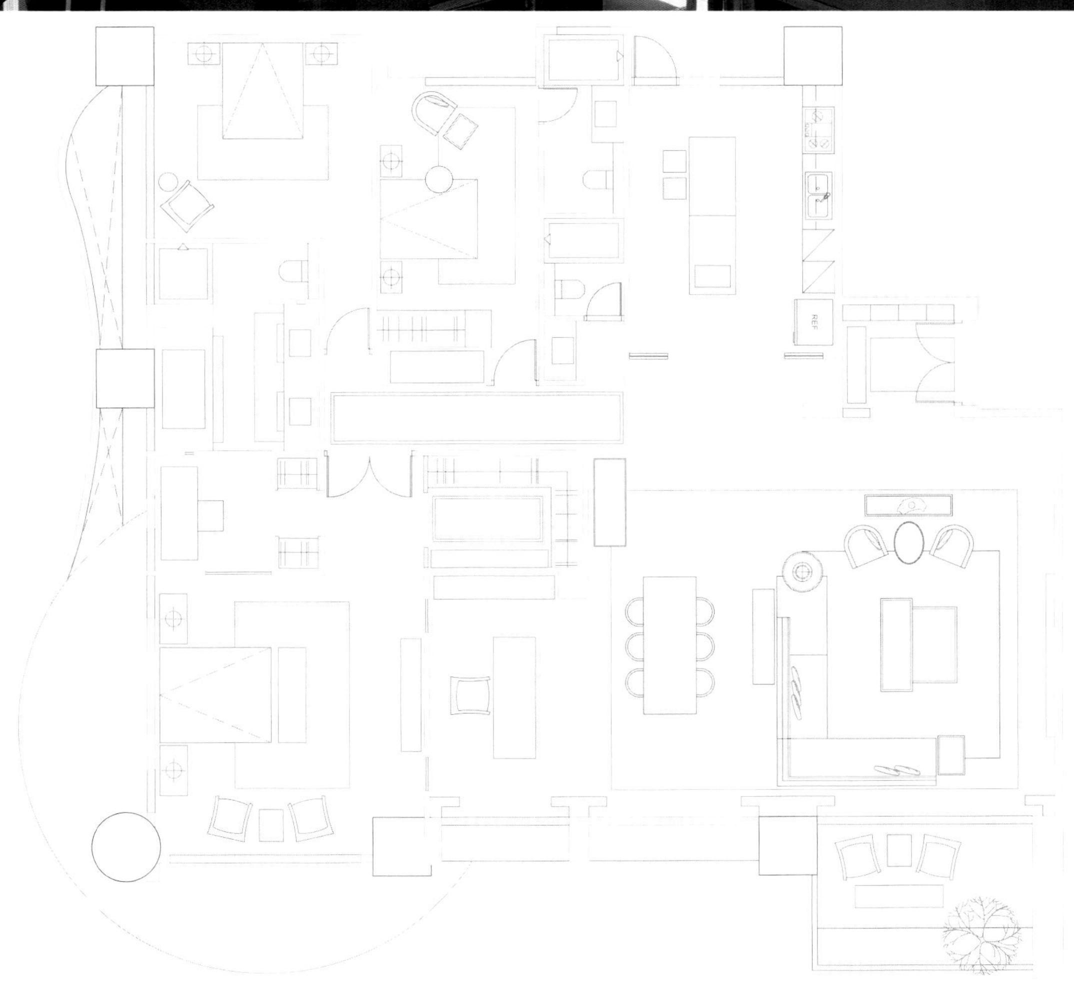
REF

DESIGN HOTELS YEAR BOOK 09
CENTURY
CENTURY
CENTURY
CENTURY
CENTURY
INTERIOR REVIEW

DESIGN HOTELS YEAR BOOK 09
DESIGN HOTELS YEAR BOOK 09
CENTURY
CENTURY
CENTURY

CENTURY

基地为一临水私人招待会所，拥有开放的河岸，自然环境条件优异，单面环河视野无限，采用大片窗引进自然光，在此可以感受到大尺度空间与自然的对话。设计决定采用“观看”的概念去思考，让空间如艺廊般感受这份难得的室内外景致。

空间规划上，以一条空间廊道向两侧展开，低姿态的配置方式强调了上段空间的流畅与下段空间的区位。墙面基调采用灰色镜面材质，呈现如水波般的自然肌理，营造出空间内部的戏剧性与感官体验。

空间中轴线的廊道空间，在将高度、宽度压缩后，凝聚成行走尺度。借由两侧艺术品的陪伴，慢慢地、静静地、渗透到其他空间，通过空间向屋主还原了原本大河的山谷景色。

MR MA'S RESIDENCE ON THE SOUTH ROAD OF DUNHUA

敦南马公馆

设计公司
玉马门创意设计有限公司

设 计 师
林厚进

项目地点
中国台湾

项目面积
152m²

主要材料
毛丝面不锈钢、赛丽石、钢构、盘多磨、木作、木纹砖

摄影师
张克智

The walls of the hallway are decorated with cultured stone so that the solid stone surface and the joint groove reflect with each other in the light, and therefore enhance the warm feeling of the space. The embedding of an ornamental cupboard produces a sense of tasteful life, but in fact, a large store room is concealed behind to satisfy the owner's needs to store anything.

After stepping in the parlour, plenty of sunshine scatters like quicksand and looks achingly beautiful. It makes different changes of light and shadow on the floor and wall as time goes by, and also brings in the sunshine to the downstairs smoothly. The furniture with the mixture of pure white and wood grain makes the space more comfortable.

The position of the top floor is fully taken advantage of. On the old roof a new ceiling is set up, and two skylights are opened up to let the natural light in so as to solve the problem of insufficient light and to make the space wider and brighter.

The simple, modest and open dining room with simple modern style professional system cabinet is suitable for both dining and business and shows the culture of being composed and reserved.

Stairs made of driftwood cedar lead to the private space downstairs – a separate study and a private studio. Large French window divides the space and lets in the light to make the space more penetrating so that the space indirectly extends to the small outdoor balcony, all of which allow the owner to enjoy a relaxing and spacious feeling while busy at work.

In the bedroom, a small balcony garden is designed specially, on the one side to improve the condition of insufficient light in the room and on the other side to protect the owner's privacy with the help of barriers and roller shutters without destroying the harmony on the whole.

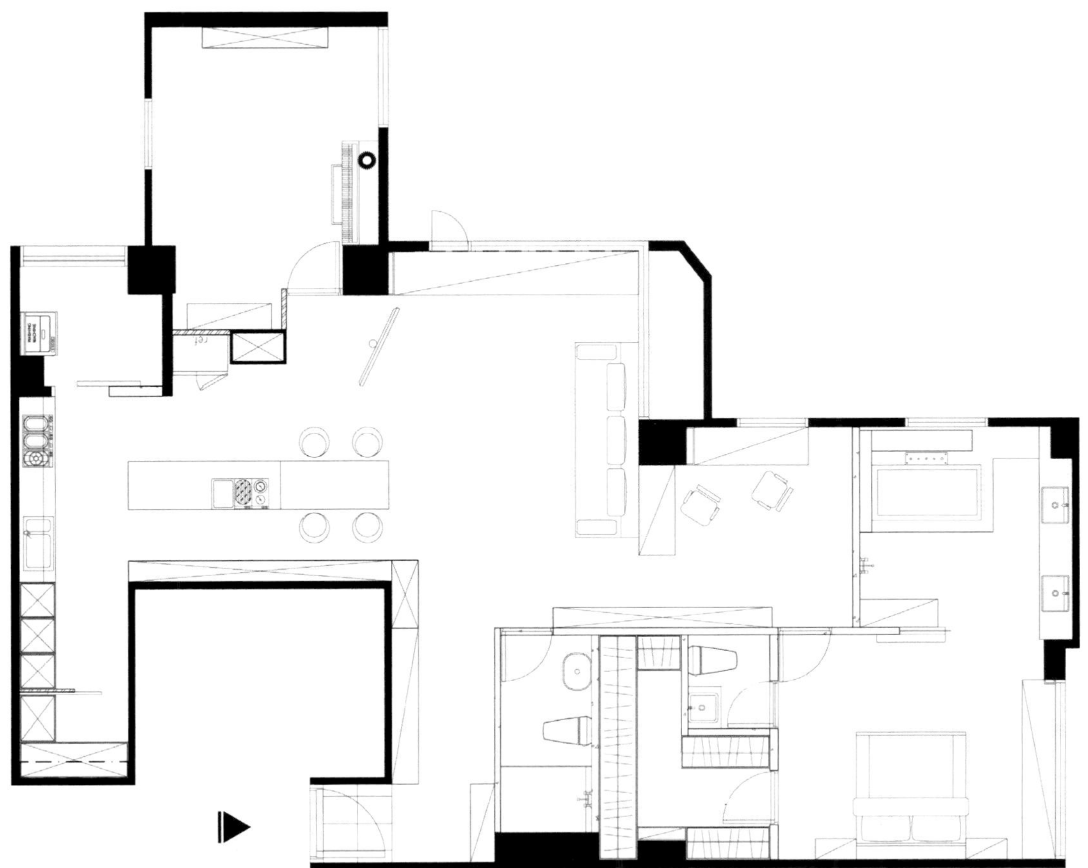

79
我們的新世界
日本史圖解

玄关处的墙面，设计师以文化石作为装饰，立体石面与接缝沟槽于光影中增加空间温度。在前方嵌入些许装饰柜点缀出生活趣味，但其实背后是一整面的储藏室，满足了屋主的收纳需求。

踏入客厅，大片日光如流沙般洒落，美丽至极。随着时间的转动在地面与墙面上营造出不同的光影变化，也顺利将阳光引入楼下空间。纯白与木纹搭配的家具，使空间看起来更加自在舒服。

利用楼顶的优势，在旧有屋顶上架设天花板，并开立了两面天窗以引进自然光，改善光线不足的问题，使空间更为明亮宽敞。

简洁而稳重的开放式餐厅，搭配现代简约的专业系统柜，用餐或洽谈公事皆可，展现沉稳与内敛的居家文化。

随着一层层浮木香杉搭配的楼梯往下来到私人空间领域——独立书房和私人工作室。以大片落地玻璃作为空间区隔，引光入室，使空间具有穿透力，间接延伸至户外小阳台，让屋主在办公忙碌的同时有轻松宽敞的感觉。

在卧房处特别打造了一座小花园阳台，利用透明玻璃窗，改善室内光线不足的问题，隔栅与卷帘仍可保护个人隐私，却不失整体和谐性。

PINZHENDI MR WU'S RESIDENCE

品臻邸吴宅

设计公司
珥本室内装修设计工程有限公司

设 计 师
陈建佑、曾耀征

项目地点
中国台湾

项目面积
182m²

主要材料
美洲薄壳胡桃木皮、雪白银狐石、深棕榄石、镀钛板、铁件

摄影师
吴启民

In the design, there is no stereotype, no strict function boundaries, no complicated construction patterns, the idea of the design is to focus on a natural and relaxing life-style.

Pecan veneer: It is a wood veneer product from US, which is used on an existing French classical style wall panel with decorative motif. The design makes use of the three-dimensional change of the panel to highlight the focus, offering a softer and elegant quality to the living room.

Silver fox stone: This material replaces the wood skirt board and applies to the French style plate as the skirt, which reveals the clear pattern of the stone and enhances the 3-D feeling. The turning point of the skirting in the living room is naturally designed as a platform for TV installation, and which in the main bedroom forms the base of the bed. In this way, the changing materials represent different functional areas, and decrease the unnecessary lines in the space.

Dark brown peridotite stone: The material is applied in the buffer area between the living room and the dining room to separate the two areas from each other. An Island styled fireplace is the dividing point between the living room and the kitchen. It links the two areas but keeps the interference away. Meanwhile, it also serves as a functional place where glasses, red wine, magazines and newspapers are stored.

Plated titanium board + painted black iron: Though the overall space is developed around the concise neo-classical lines, the designers also want to add modern and urban elements to the project to make the fashionable materials and styles contrast with each other to present a certain kind of fun and conflict. This method is applied to the choice of furniture as well – diversified combination of furniture styles creates a gentle and reserved neo-classical effect.

Untouched Ali Hanan &Pip Norris
ANDREW MARTIN INTERIOR DESIGN REVIEW

这里没有规矩的形式，没有严谨的机能，也没有繁复的样貌，设计师塑造的是生活当中自然自在的重量。

美洲薄壳胡桃木皮：运用于法式壁板的面材，利用线板的立体变化来凸显立面设计的重点，并使空间线条多了典雅的温柔感。

雪白银狐石：运用于法式壁板的踢脚，以石材取代木头烤漆踢脚板，让石材的晶透与纹路增强线板的层次感。在客厅的踢脚转折处自然形成放置电视的平台，主卧室则形成起居使用的卧榻底座，以材料的延伸来转换机能，从而减低不必要的空间线条。

深棕榄石：运用于区隔客厅与餐厅的中介空间。中岛式的壁炉柜是客厅与餐厅的分界点，让二者互相关联又不会互相打扰。同时，它也是一个机能性的服务空间，提供了酒杯与红酒、报章杂志的收纳功能。

镀钛板＋黑铁烤漆：虽然整体空间以简洁的新古典线条为主轴，但我们仍希望为空间加入现代感的都会元素，让材料与样式在冲突中产生趣味与对比。这样的搭配也体现在家具的选择上，让多样性的家具样式相互组合，呈现出低调与内敛的氛围。

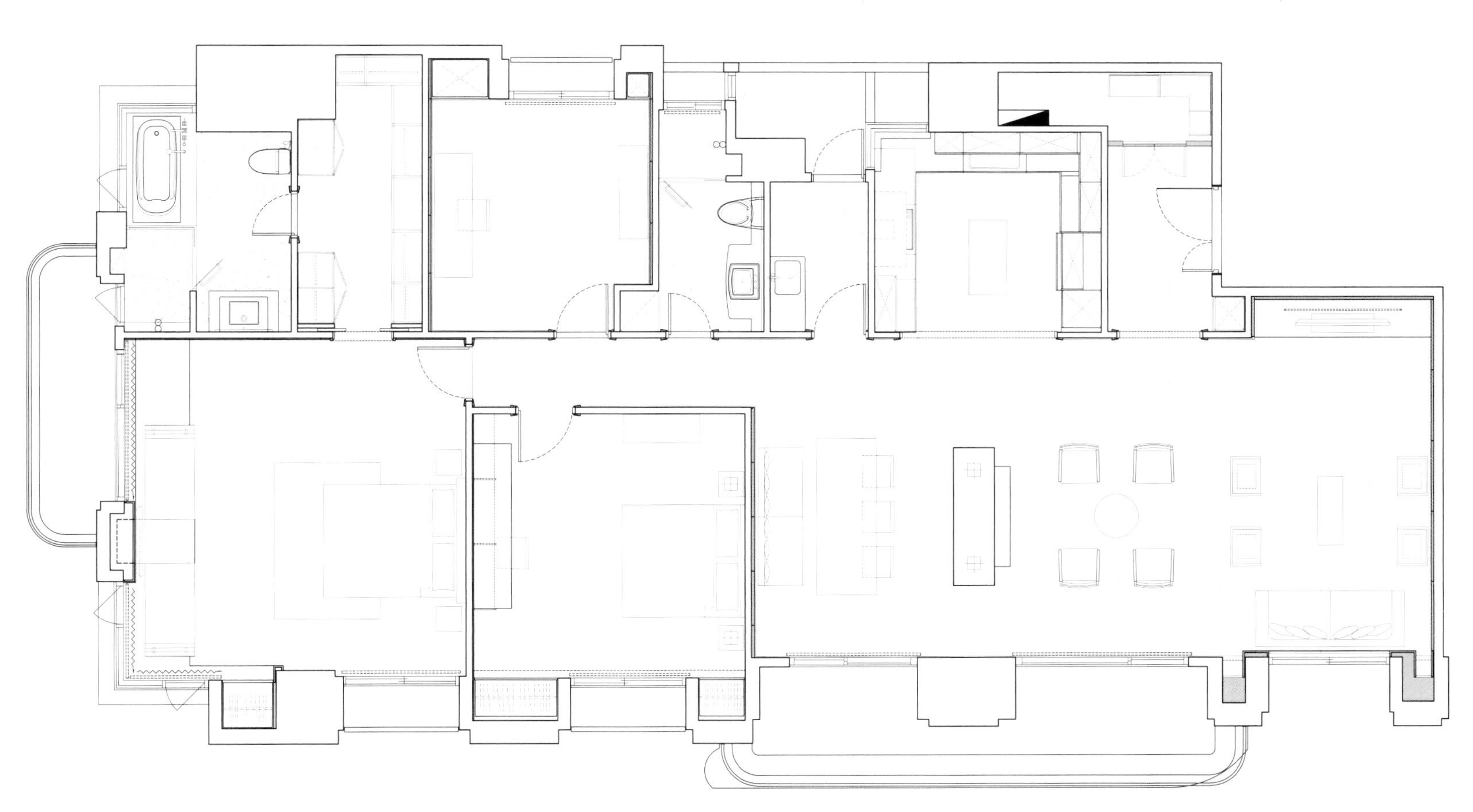

CENTURY

china LIVING
IN CHINA

MR LIN'S RESIDENCE IN LINKOU

林口林公馆

设计公司
玉马门创意设计有限公司

设 计 师
林厚进

项目地点
中国台湾

项目面积
396m^2

主要材料
超耐磨地板、钢材、松木、结晶水泥、陶瓷烤漆、马赛克砖

摄影师
张克智

The owner of the residence, who is engaged in industrial design, has a high requirement for accuracy and balanced proportion of details, so the design is based on the concept of "less is more", which is in accordance with the owner's character as well as the open and concise design style.

The first floor includes a dining room, a kitchen and a living room. White ceiling decorated only with lighting tracks is consistent with the space constructed with the pure white surface. An L-shaped full-scaled window generously invites abundant light into the inner space, stretching the light onto the wall in the dining room due to the skillfully designed gradation board around the windows. Though it is a non-blocked design, the hallway is created by the outline of the couch, and thus smartly divides the space for different functions. An LCD-screened TV set adjustable to different angles is hung on a light steel column. In this way, a simple design is presented with simple lines, echoed with the large area of white surface which changes with the change of light and shadow to form a sharp contrast, and the asymmetrically hanged TV is a balance of the whole space.

The natural light changes the vision, enlarges the space in every level and creates a dynamic shadow of the grids in the study on the second floor. The kids' room is on the third floor and the design tries to create a bright and energetic place with Naples yellow and white colors. The dressing room is on the fourth floor with open windows overhead to not only let in the sunlight, but as a way to ventilate, in order to get rid of the dampness and darkness. The terrace on the top floor is painted with Laim green, matching perfectly with the pine material to create a bold and relaxing atmosphere. Abundant sunlight and bright colors inject energy into every corner of the whole space here.

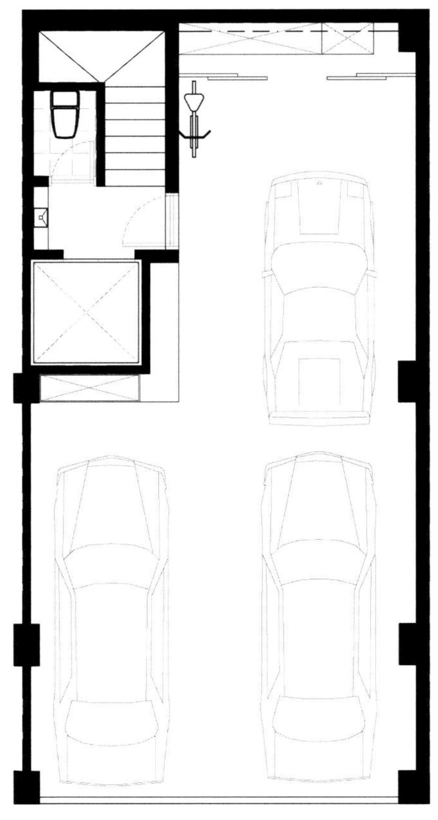

B1F FLOOR PLAN
B1 楼平面图

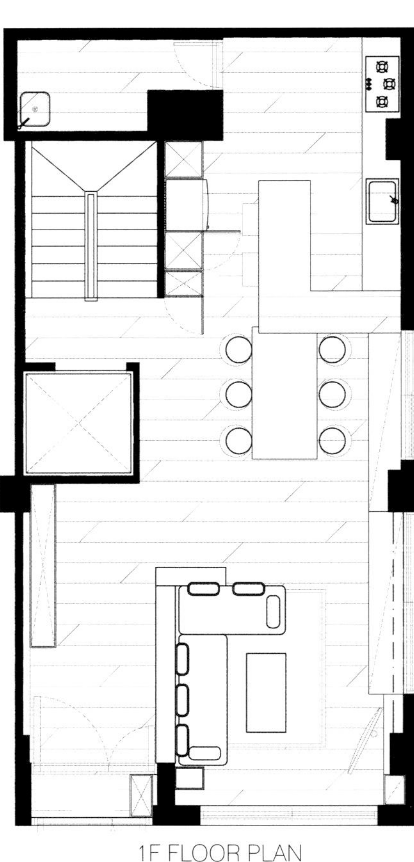

1F FLOOR PLAN
一楼平面图

从事工业设计的屋主对于设计细节的精准度与平衡比例相当讲究，因此设计师以“减法”做为设计根基，贴近屋主的个性及所要求的“开放简洁”。

一楼为连贯餐厨和客厅的开放空间，在洁白且不加修饰的天花板上只见照明轨道顺延牵引，并以同色系的漆面塑造空间的一致性。门口处L形的大面采光设计、全面式透明采光窗到窗边的层次台面设计，结合生活机能，将光线由外向内延伸至餐厅墙面。虽然是无阻隔设计，利用沙发线条圈出玄关意象，巧妙区分了空间。轻型钢柱上悬放可调整角度的薄形电视，让空间只见轻简的设计线条，随着光影流动变化以大片的留白作为陪衬，最后以非对称悬挂的电视平衡空间。

光线照射使空间视觉转换，放大了每层空间的面积，因此也在二楼书房格栅上映像出时序的阴影。三楼小孩房，设计师以鲜明的鹅黄色佐以纯白色铺陈明亮与活力感。四楼更衣室上方开立小窗，除了让阳光透射入内外，还可借着光热化解更衣室的湿气与黝暗。最顶楼的露台处以清爽的莱姆绿墙漆搭配松木，营造粗犷与悠闲感，借着日光与明亮色彩的组合，带动了每个空间的活力。

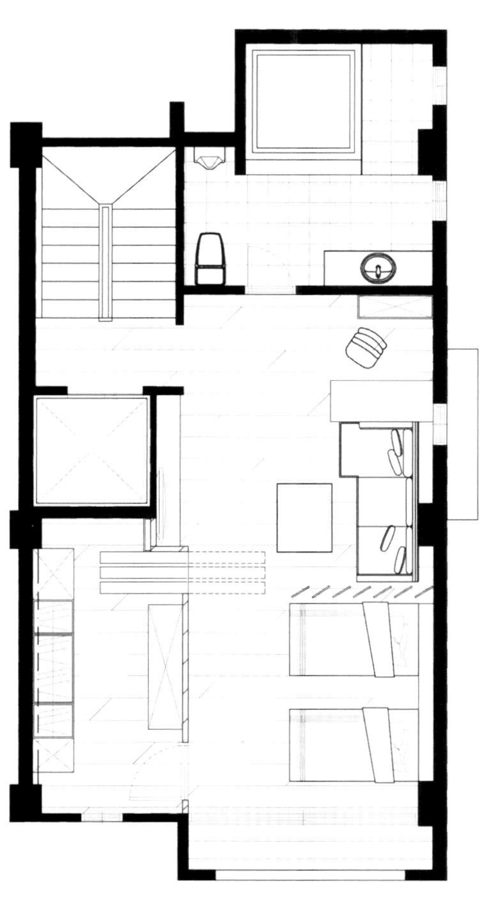

2F FLOOR PLAN
二楼平面图

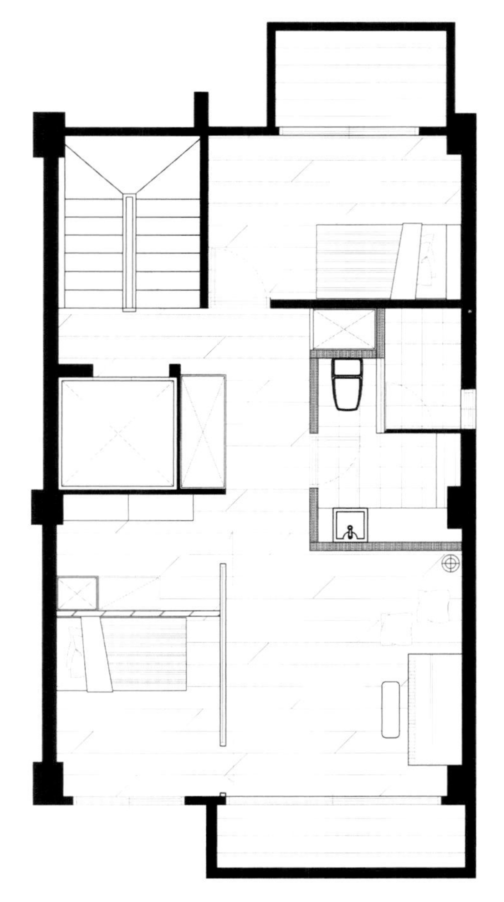

3F FLOOR PLAN
三楼平面图

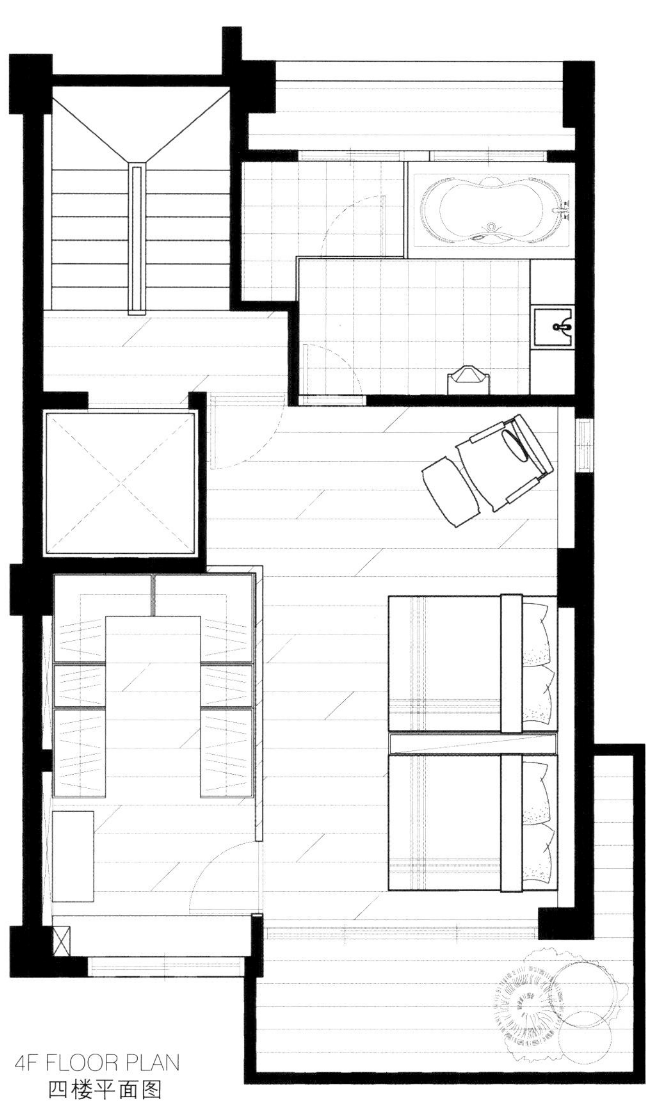

4F FLOOR PLAN
四楼平面图

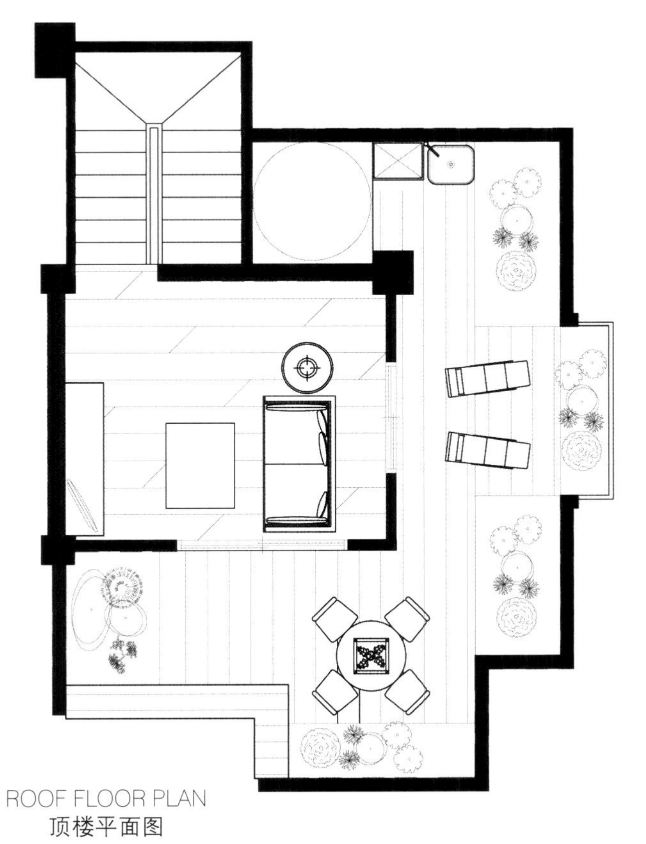

ROOF FLOOR PLAN
顶楼平面图

L HOUSE

L宅

设计公司
城市设计

设 计 师
陈连武

项目地点
中国台湾

项目面积
353m^2

The client is a big fan of manual art crafts and antiques, who deeply loves the natural texture and smooth surface of materials. Therefore, the designer uses curio shelves as original model for the wall designing to display valuable antiques. Various kinds of L-shaped factors are applied to the walls in the space. Besides, the ceiling in the opening area on the first floor that is interlaced with wooden boards is presented via a traditional technology, therefore, air conditioning system, lights and other modern equipments are hidden from sight. The DAIKIN concealed air conditioner hanged on the ceiling integrates with the overall design perfectly.

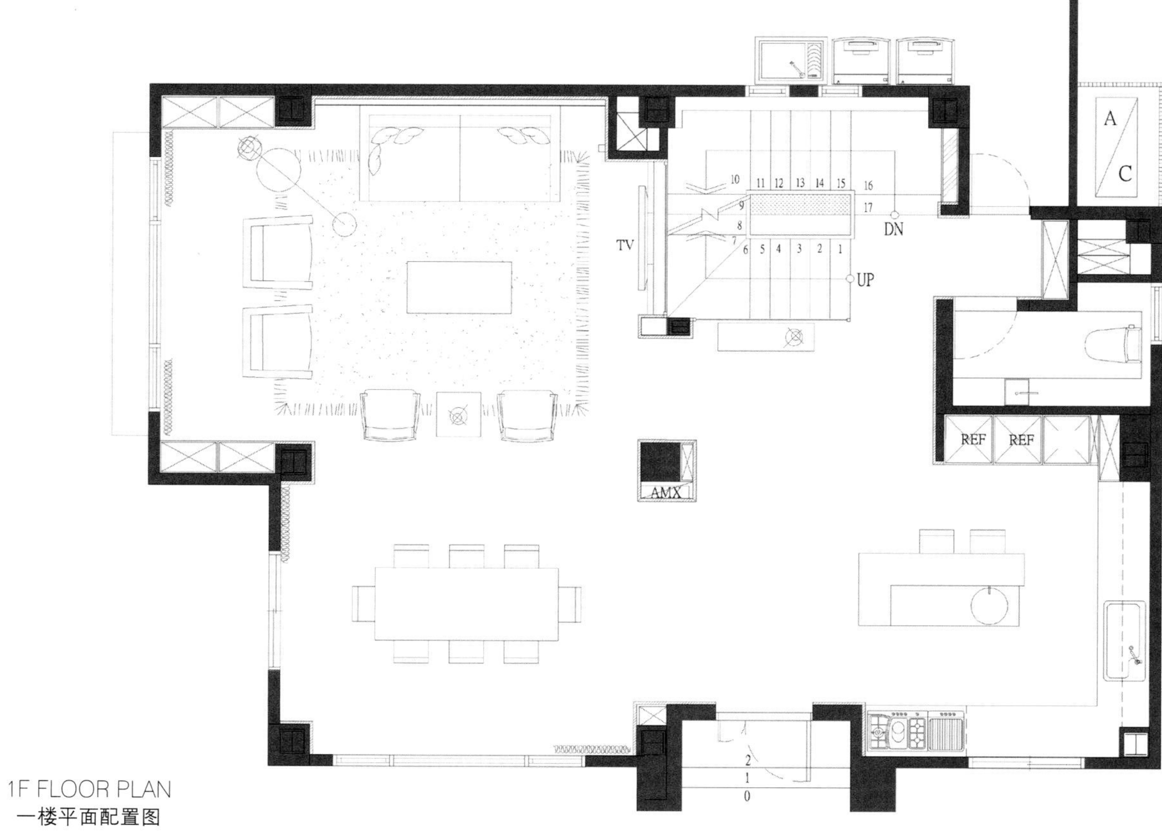

1F FLOOR PLAN
一楼平面配置图

喜爱材料自然肌理与光泽的屋主，有收藏中式手工艺品与古玩的爱好。因此，设计师在立面的设计语汇上，以陈列古玩使用的博古架为原型，将L形变化后的各种形态，表现在各个空间立面上。另外在天花板平面的设计上，一楼的开放空间采用木夹板编织的方式，用传统手工艺的手法表现，将空调、灯具等现代设备包覆其中。大金冷气的吊隐式出回风系统将本案天花造型整合为一。

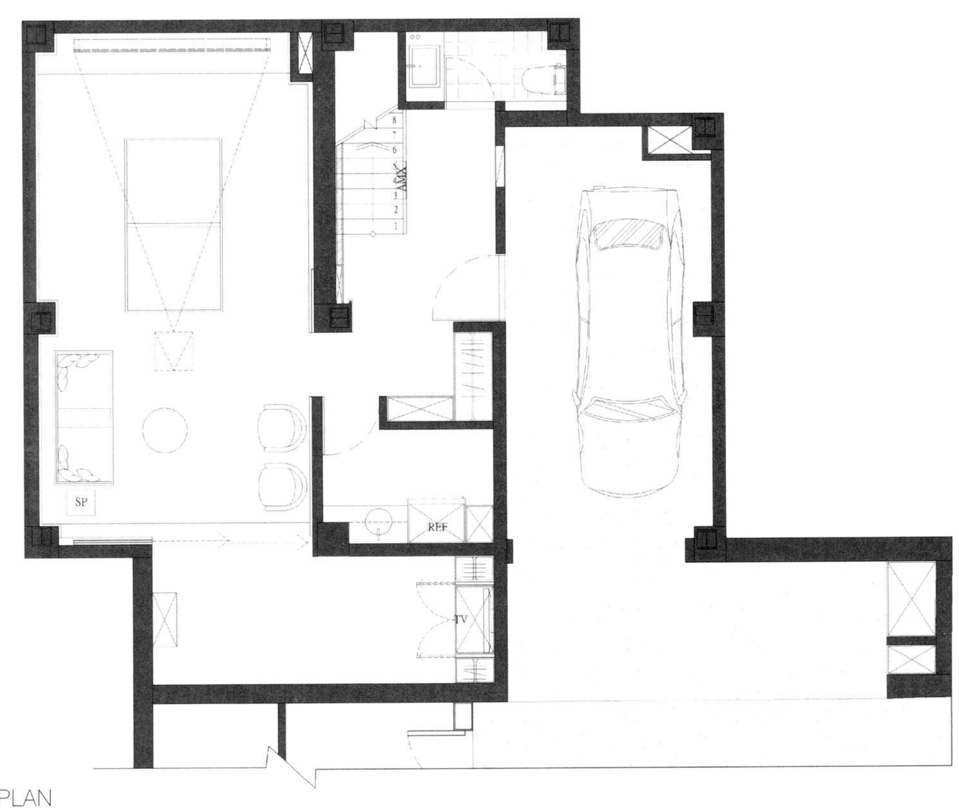

2F FLOOR PLAN
二楼平面配置图

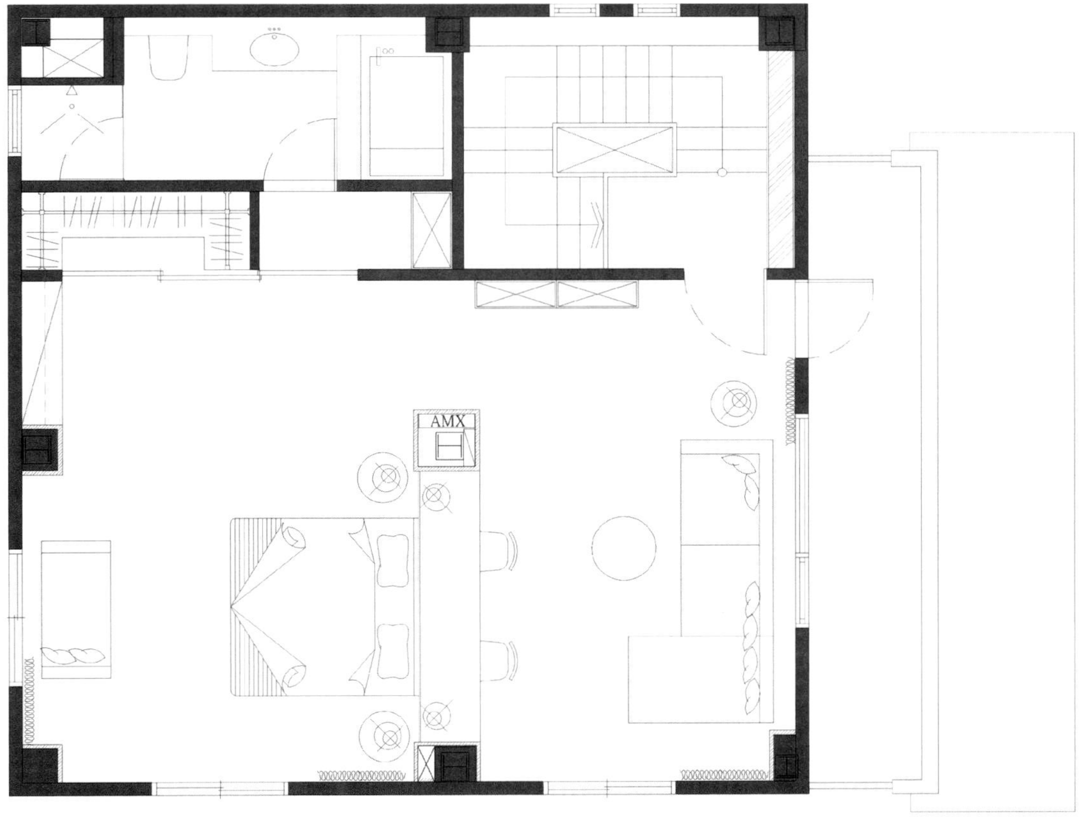

3F FLOOR PLAN
三楼平面配置图

SAMPLE RESIDENCE OF YUANJIANZHU IN ZHUBEI

竹北原见筑样板房

设计公司
陆希杰设计事业有限公司

设 计 师
陆希杰

参与设计
刘冠汉、张琼之、林丽纹

项目地点
中国台湾

项目面积
172m^2

主要材料
盘多磨、南方松、卡拉拉白、栓木、马赛克、钢琴烤漆、乱纹不锈钢

摄影师
Marc Gerritsen

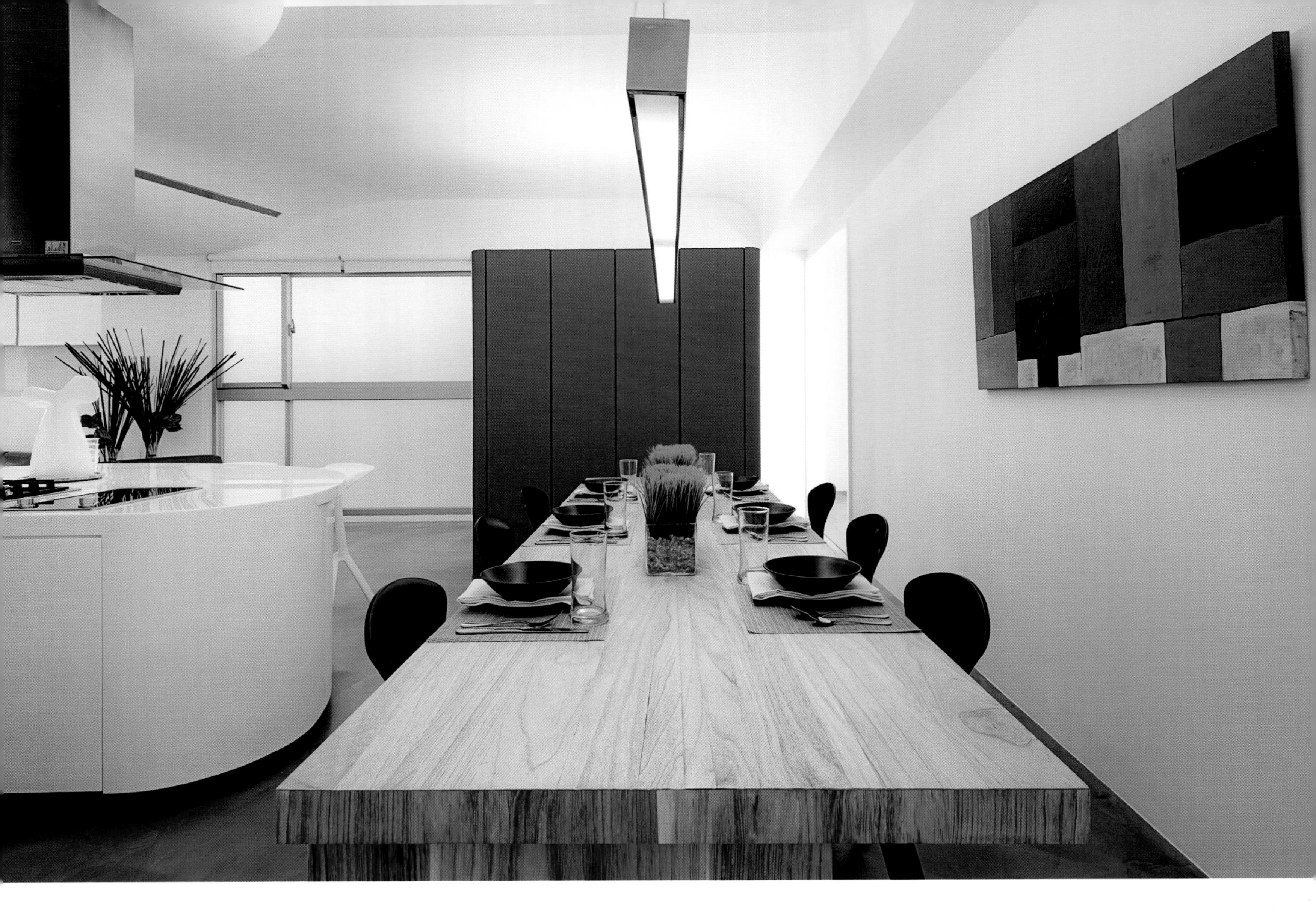

Sample residence of Yuanjianzhu is designed by famous Japanese architect Hiroshi Hara. The concept of this project is to make two buildings surround a courtyard to form a space similar to a courtyard. In the building, every household has a two-storey high terrace.

The inside decoration is in high correspondence with the design of the building. The high terrace is the focus of interior space, the study with a terrace is the first thing one will pay attention to after entering from the hallway. Continuous active lines run through all the open and free flat surfaces so that every corner in the space can be seen in different perspectives. Limited by the foundation of the building, light in the space is not adequate and thus, it is relatively private. But the open style design enables the inside to present a bright and open visual effect.

Every unit of the building is divided into separated area. The pattern of arrangement looks continuous but in fact separated. Once the glass door to the terrace is opened, the space and atmosphere of the living room are extended. The terrace that serves as outdoor study echoes with the kid's playing area. The grass-green built-in cabinet in the kid's playing area makes the space more natural. On the ceiling, curve lines are used to decorate the beams. The ceiling itself is a light with soft lights emitting from changing and twisting slits to make the whole space natural and integrated.

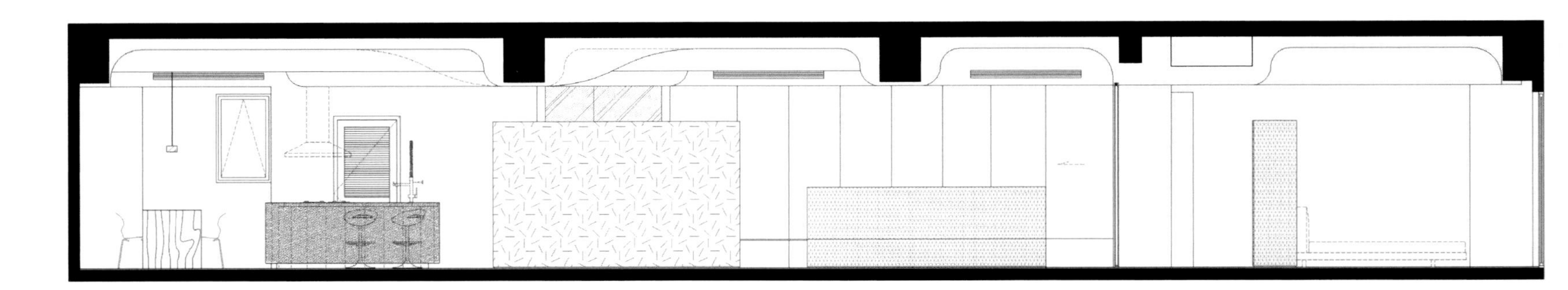

0 1 2 3 (M)

原见筑样板房的建筑规划者是日本建筑师原广司，其设计概念是利用两栋建筑环绕着中庭的方式，来形成一个类似合院的空间，并且每一户都拥有两层楼高的挑高阳台。

本设计案以延续建筑的精神来执行室内的规划。其中，室内空间的部分以挑高的阳台为焦点，从玄关进入后的第一印象即为阳台书房，接着就是以连续的动线贯穿整个开放的自由平面，让视觉角度可以穿透到空间中的每个角落。此空间由于地基的条件关系，较少有对外的大面采光，空间产生了较隐蔽的特性，但是因为开放的配置，使得内部设计开阔、明亮。

每个单元都有各自独立的区域划分，看似连续的格局安排，却又是明确的个体。阳台的玻璃门窗拉开后，客厅的氛围便随之延伸，阳台作为室外的书房与室内小朋友使用的游戏区相互对应，室内小朋友使用的游戏区的草绿色置物墙柜衬托出由外而内延伸的自然语境。天花板利用弧形的线条来修饰梁柱，而天花板本身就是一个灯具，柔和的光线从间接的转折中流露出来，高低起伏地变化着，让空间自然融合。

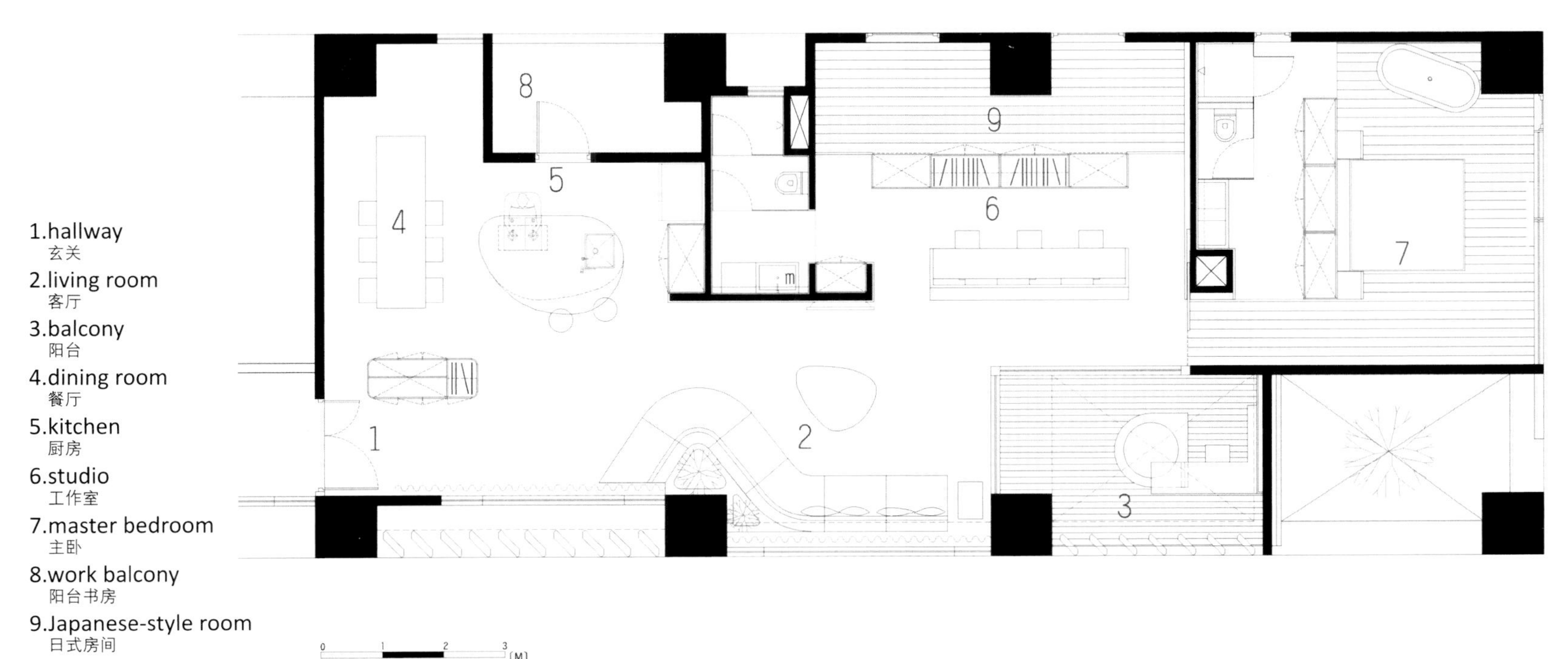
1.hallway
玄关
2.living room
客厅
3.balcony
阳台
4.dining room
餐厅
5.kitchen
厨房
6.studio
工作室
7.master bedroom
主卧
8.work balcony
阳台书房
9.Japanese-style room
日式房间
0 1 2 3 (M)

LINKOU WORLD CHIEF SHOW FLAT

林口世界首席样品屋

设计公司
动象国际室内装修有限公司

设 计 师
谭精忠

参与设计
陈姿蓉、许玉臻

项目地点
中国台湾

项目面积
172m²

主要材料
实木钢刷木皮板染色、喷漆、镀钛、壁布、橡木染色地板、灰网石、雕刻白、皮革、夹纱玻璃、灰镜、手工贝壳

From the entrance area, people can enjoy a view about 10 meters wide, which serves fully as a representative of the grand and luxurious mansion. Dark grey marble paved on the floor in the common areas is in sharp contrast with the white painted walls, and the designers create a strong sense of calmness and reservation. The regularly segmentation of the wall and dark steel-like painted wood not only divide the space gently but also enhance the tense of the space arrangement.

The living room includes a reading area which prolongs the visual effect of the space. The high class cabinet in the reading area with titanium metal shelves and dark steel-like painted wood presented with nice furniture becomes a view enjoyed from the living room.

All the cabinets in the living room and the dining area are designed in an invisible way in which the effect of the wall segmentation is well extended. Besides, the cabinet behind the wall provides a clear outline of the overall space and enables a high capability of storage as well. The paintings hung on the white painted wall not only show the owner's elegance and refined taste, but also strengthen the effect of the combination of art and residence.

The kitchen is separated from other areas by a gauze-glass screen door, in which a clear and bright visual effect is achieved by using the white filmed glass as the ceiling illumination.

The constant concept of "fashion" is also applied to the main bedroom which is accommodated with all facilities in high-ranking hotels. The washing room area and the sleeping area are in an open design which enlarges the space of the entire bedroom. In terms of material, it also adopts dark steel-like painted wood in contrast with the white carved stone, so the strong sense of calmness and reservation is fully deployed in this space, too.

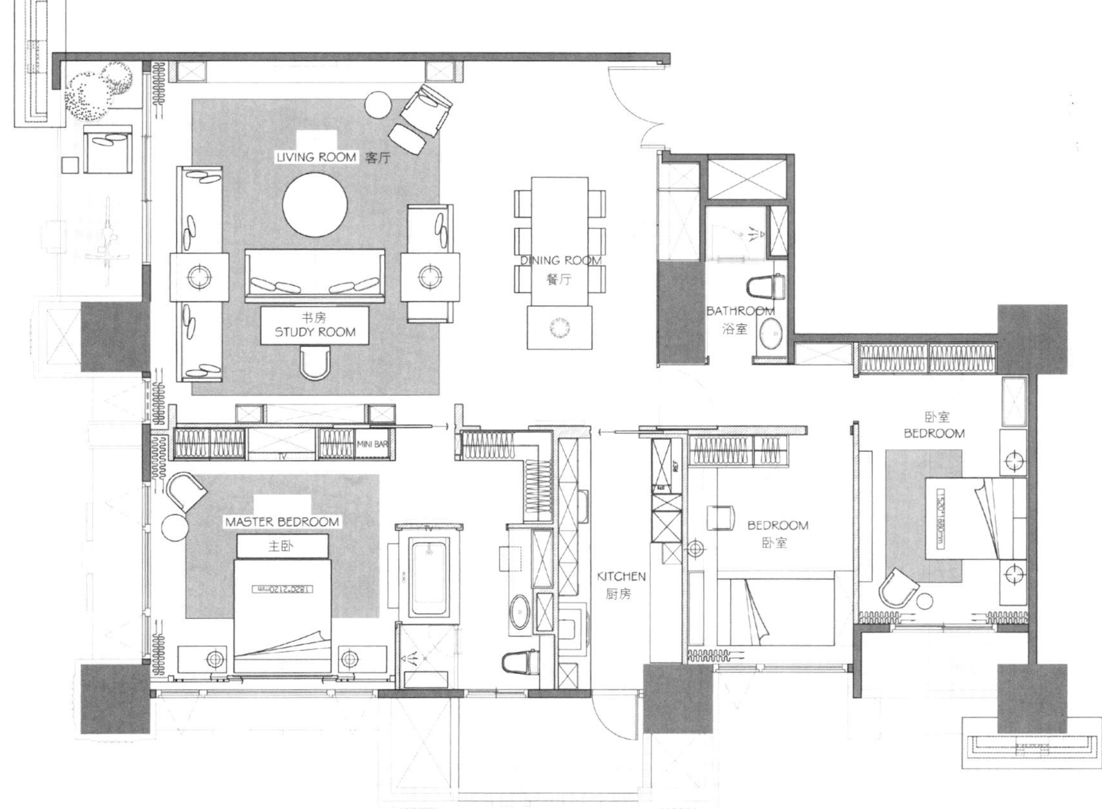

进门时，10 米宽的空间视野，呈现出大器的豪宅气势，公共空间地面铺设灰黑色的大理石石材，与白色喷漆壁面形成强烈的对比，进而衬托出空间的沉稳与内敛。墙面规律性的分割与染深色的钢刷木皮，不仅将空间区域以软性手法加以分界，更加凸显了空间的张力。

客厅结合阅读区，不仅拉大了空间的视觉效果，同时，阅读区展示柜体的处理上，以镀钛金属框架结合深色钢刷木皮，并以精品家具呈现出高质感的展示柜，使其成为客厅区域的端景。

客厅、餐厅区域的柜体均使用暗柜方式处理，延续了墙面的分割效果。同时，隐藏在墙面后方的柜体，除使整体空间线条干净外，更提供了高机能的收纳空间。素白的喷漆墙面吊饰画作艺术品，衬托出空间的优雅与品位，强化了住宅与艺术结合后的加分效果。

餐厅端景墙的夹纱门片为厨房区域，天花板照明设计选用白膜玻璃，使厨房空间更明亮。

主卧室的设计延续一贯的时尚风格，以高级饭店配置为取向。厕所与卧房区将开放式效果作为配置理念，不仅拉大了主卧房的整体空间感受，在材质选用上，继续使用了客厅的深色钢刷木皮材质，衬上厕所以白色为基底的雕刻白石材，续写了沉稳与内敛的质感。

LOST SPACE

消失的空间

设计公司
台湾建构线设计有限公司

设 计 师
沈志忠

项目地点
中国台湾

项目面积
233m²

Spatial interfaces are redefined to create a flexible layout; "changeable" is the theme of design. The living room, study and kitchen are all separated by low walls that can be moved, the TV cabinet can be rotated to face the study or kitchen. A blue screen that creates a virtual interface expresses the relationship between spatial boundaries, achieving a smooth line of motion and a coherent view. In addition, there is a hidden sliding door that can turn the virtual boundary into a solid boundary.

The color tone of the space uses white as a base, and the designer proportional divides materials and colors to turn colors into a visual image. The open kitchen creates a rich visual experience with its purple and blue shade, and the lights set off a bold and fashionable match of space with colors.

When you enter the master bedroom, the color of the wooden skin creates a natural shade, and sets off the rich layers of materials when put in contrast with the simple arrangement of white marble; therefore the space is low key and warm. Follow the steps inside the master bathroom to the raised shower room, and it shifts your view to the vertical axis, where you will be able to feel the energy and smooth flow of the space.

为了创造弹性的使用格局，空间界面被重新定义，“可变动”是此空间的设计主轴。客厅、书房、厨房皆用可移动的矮墙作为空间区隔，旋转的电视柜随着角度的移动，可转折到书房与厨房两个不同性质的空间中。蓝色隔屏勾勒出虚性接口，表达出空间的界线关系，让行走动线与视觉流畅。隐藏式拉门亦可把虚的界线转换成实的隔间。

空间的色调以白色为基底，材质、色彩按比例分割，让色彩变成一种视觉景象。开放式厨房在紫色、蓝色调中，为空间注入了丰富的视觉体验。在灯光的调和下，映衬出空间与色彩前卫大胆的搭配。

进到主卧室，木皮本身的色泽让色调回归自然，显得低调而温润，既与白色大理石的冷冽简约形成绝对的对比，又衬托出材质的丰富层次。主浴室顺着阶梯，踏上抬高的淋浴间，设计把视觉的层次拉到垂直轴，产生一种活泼的流畅感。

inSIDE : outSIDE
LUXURY HOTELS
CODEC DESIGN
VIDEO COMPRESSION

AMERICAN STYLE SUNSHINE RESIDENCE

美式日光宅

设计公司
常玉国际设计工程有限公司

设 计 师
吴衍霖、许秀娜

项目地点
中国台湾

项目面积
165m²

主要材料
银白玉大理石、板岩、进口壁纸、烤漆、大理石、刻花、大理石拼花、人造石台面、皮革、裱布、茶镜

Designers rearrange the layout inside the residence by injecting the white and concise American style the owner expected to scheme out areas including the dining room, the main bedroom and the living room, in which the owner can enjoy the great sunshine.

The hallway is decorated with beautiful potted flowers put on a poppo metal table with carved designs. The white grids filter intense sunlight, the soft sunlight gets into the room through the clear glass creating a harmonious scene with the light and shadow from the French window in the living room. Furthermore, the tailored marble TV wall and unique furniture in different shapes and patterns together create a vivid modern American style of life.

Behind the white jade marble TV wall is the open kitchen. Two angles of illumination have been reserved on the one side with the clear glass enlarging the visual effect of the space and on the other side with Roman curtain adjusting the intense of sunlight. Beige curtains are hung under the transparent ceilings, and a cabinet in the image of a fireplace stands in front of the yellow slate wall. The curving lines of the curtains combined with the naturalness and simplicity of the materials allowing the owner to enjoy the involvement with nature to the most.

In the main bedroom, the imported wall paper that is in contrast with the leather bedside enhances the sense of lines under the sunlight. White carved flowers on both sides of the bed soften the tense of the view. Beside the window, a love seat sofa is placed in the best lighting area, where the owner can enjoy reading in the afternoon in the American style terrace.

Apart from the accurately calculated storage space, the designers leave proper blank space for the owner to create his own way of life and to enjoy the cozy feeling of a comfortable vacation in the American styled lines.

设计师重新调整室内格局，融入屋主期待的白色简洁美式风格，规划出可享受日光照耀的餐厅与主卧、起居室。

玄关处美丽的盆花摆放在普普金属刻花的端景桌上，白色格栅挡住了强烈的光线，穿透清玻璃的光折射出柔和的光晕，与来自客厅落地窗外的光影合二为一，搭配设计师量身规划的大理石电视墙及奇趣各异的造型家具，交织出一幅动人的现代美式风范。

银白玉大理石电视墙后方是开放设计的厨房空间，设计师保留两个角度的光照，借由清玻璃放大空间视觉，并以罗马帘调整光谱的强度。米黄色的布幔吊挂于透明设计的天花板下方，壁炉意象的展示柜立于黄色板岩墙面前，布幔悬垂的弧度加上自然质朴的建材选择，使屋主在温暖的氛围里享受与自然相伴的生活情调。

在光线的照耀下，主卧室的进口壁纸与皮革床头绷布搭构出强烈的线条感，床头两侧则辅以白色浮雕刻花软化视觉张力，另在采光最好的临窗区摆放了舒适的双人沙发，屋主可在午后坐在美式露台上享受惬意的时光。

设计师在精准地计算出需求的收纳量之后，适当留白，为屋主预留出挥洒的空间，在美式线条里体现温暖度假的美好。

ZEN STYLE HOUSING
禅风居家

设计公司
玳尔室内设计有限公司

设 计 师
朱志峰

项目面积
264m^2

主要材料
木皮染黑、银箔、铝框、铁件、天然石材、烤漆玻璃、海岛型木地板

+200
STO

The designer expertly merges the gentle, implicit and quiet features of Japanese Zen style in the whole space. In terms of the choice of lines and furniture, luxurious neo-classical furniture and the lines of modern grille are joined together to highlight the visual focus of European style and Japanese Zen style to create a harmonious and spacious atmosphere.

The layout is slightly adapted so as to make the active line more consistent with the occupier's life style which can be seen clearly from the boundary lines of the space. The living room is in a direct connection to the dining room and the study, with only iron screen or glass sliding doors as partitions, making the space brighter and more transparent. The lines formed by silver foil in the living room extend and wind into the dining room, pass through the sliding door and the display counter, and then stop, to link two areas of space naturally together. Because of the owner's habit of having light meals, the kitchen and the dining room are arranged in the same space, only divided by a sliding door. In the dining area, annular lighting on the ceiling corresponds well with the round glass table, and the stable dark facades integrated with the lines and frames in the same proportions match perfectly with the design emphasis in the hallway. Thanks to the corresponding design in details, the whole space is filled with a taste of vivid life.

The master's room is simple but luxurious, with grilles and lines in Zen style and wall paper with classical totem on. The multi-functional video wall on one side is the highlight of the room. Delicately selected bevel shutters ensure the transparency and privacy of the sleeping area and the locker room. The sleeping area becomes more simple and functional with TV set moved to one side of the room.

设计师将日式禅风较为温润、含蓄、宁静的特质，表现于总体的情境氛围中。在线条、家具的选择上，巧妙地运用新古典的华丽家居、现代化的格栅，使欧式风格和日式禅风汇整成为视觉主题，和谐共存于宽敞的空间之中。

设计略微调整了居家格局，使动线更为贴近居住者的生活习惯，这一点在空间界线的定义上体现得尤为明显。客厅直接连结餐厅、书房，铁件屏风及玻璃拉门的区隔，使空间更为通透宽敞。沿着客厅的银箔线性引导，转折进入餐厅，越过玻璃拉门、展示柜仍向前延伸一格，作为两个空间的连结元素。为呼应居住者鲜少开火的轻食习惯，仅以拉门为分界，将厨房、餐厅整合在同一个大空间内。环视用餐区域，环状灯饰对应玻璃圆桌，稳重深色立面融入等比例的线条框架，对应玄关的设计重点，使空间充满前呼后应的生活情趣。

简洁中带有华丽度的主卧室，融入禅风印象的格栅线条、古典图腾壁纸，侧面影音墙的多功能性亦成为设计的重点。特意选用斜面的百叶素材，调整睡眠区至衣物间双向应有的穿透感及隐蔽性。将电视移至侧面的做法，也让睡眠区域更加单纯化。

BIG HUG

设计公司
近境制作

设计师
唐忠汉

项目面积
139m²

主要材料
橡木钢刷、铁件、茶玻、石材、玻璃烤漆

CHUANPU
川普

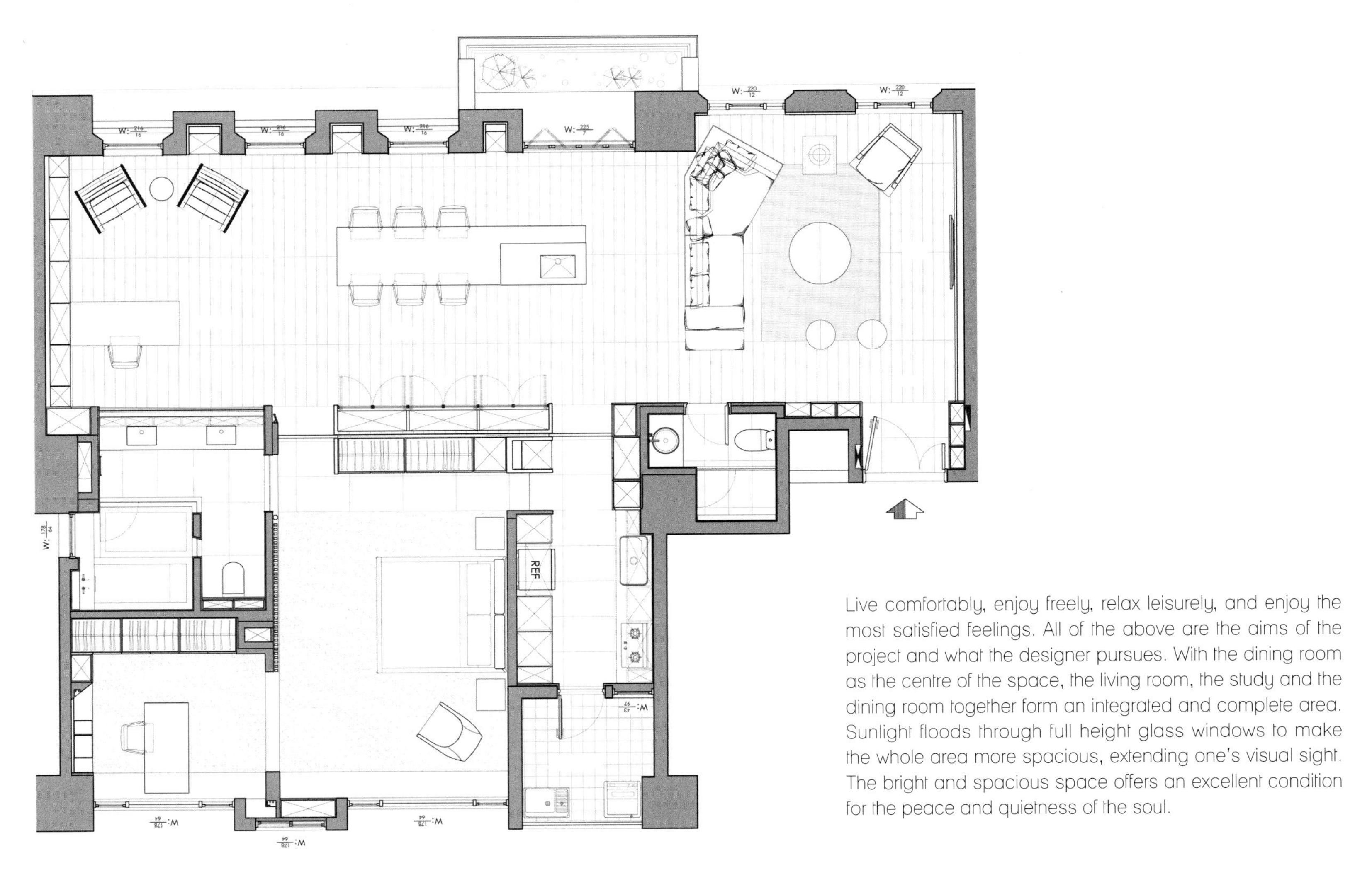

Live comfortably, enjoy freely, relax leisurely, and enjoy the most satisfied feelings. All of the above are the aims of the project and what the designer pursues. With the dining room as the centre of the space, the living room, the study and the dining room together form an integrated and complete area. Sunlight floods through full height glass windows to make the whole area more spacious, extending one's visual sight. The bright and spacious space offers an excellent condition for the peace and quietness of the soul.

让屋主用最舒服的方式生活，用最自在的角度享受，用最从容的状态放松，用最满足的情感享受是设计师设计此案的追求。空间的重心，由餐厅空间发散开来，客厅、书房与餐厅空间形成一个完整的区块。阳光透过连续的长窗洒落满地，形成视觉上的延伸，放大了空间的尺度。敞亮的空间为心灵的平和静谧创造了绝佳的条件。

GENE NEW ORIENT

捷年新东方

设计公司
近境制作

设 计 师
唐忠汉

项目面积
165m²

主要材料
洞石、柚木集成材、烟熏橡木、铁件

The usage of materials creates different levels of space: the turning point of the ceiling and wall provides an open and common area which escapes from the regional framework; different floor materials have been used as boundaries of areas.

The wall in the hallway made from handmade unburnt earth triggers visitors' intellectual and literal affections, from which people can enjoy a non-blocked view of the open and continuous space including the living room, the dining room, and the bar area. There are entrances on both sides between the dining room and the study. Through the clear glass, one can get a penetrating view. White tiles used in the space not only infuse a sense of smart into the study, but also form a sharp contrast with the city sights outside the window.

Various plans are made on the original living space, the common areas are designed as open communication space which can be shared with others freely. Dark colors used in the design create a strong sense of humanity. Black elements are used in the dark-color-toned design, and the stable and classical feeling is presented by the large amount usage of classical teakwood and "T-shaped" grain, together with the use of mirror, creating an elegant and sober living atmosphere.

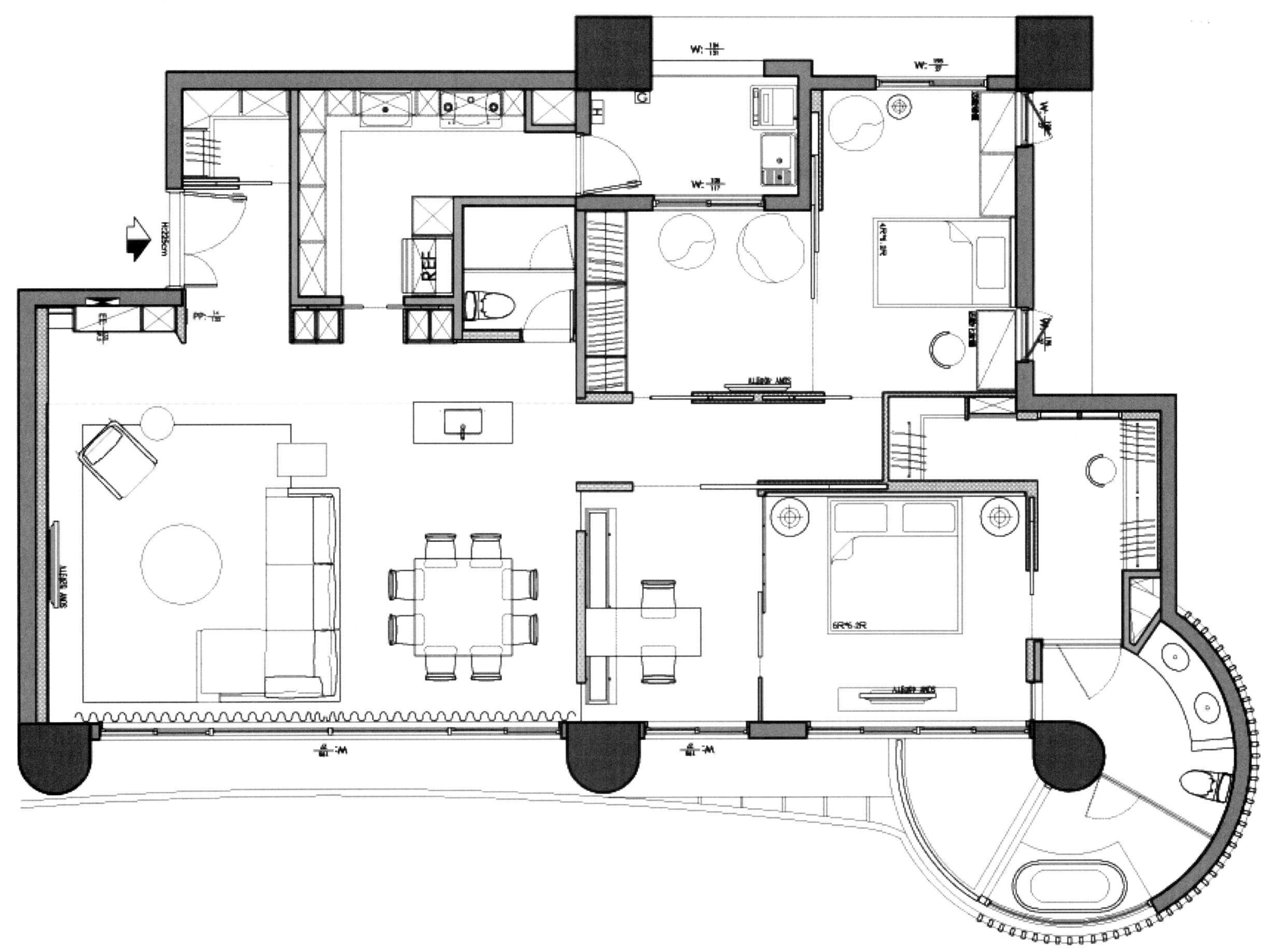

利用材质体现空间层次：天花和地壁的转折处理作出开放式的公共空间，跳脱了区域的框架；地面也通过材质的转换限定了不同的区域。

玄关处手工坯土制作的墙面，体现了人文知性的情感，又不阻碍视觉的穿透及延展，客厅、餐厅、吧台区整合成开放式厅区。餐厅与书房之间，保留左右端开口，以清玻璃诠释通透景况，白砖意象给予书房利落的质感，更衬托出窗框外的城市风光。

在既有的生活空间里做出不同的规划，将公共区域设计成开阔的交谊空间，能够自由自在地分享空间。将深色元素渗透至作品中，营造出深刻的人文温度。在深色主调的原则里投入了大量的黑色元素，空间大量运用柚木的复古色调及其山形纹，呈现了沉稳而复古的感情，与镜面材质完整地凝聚了空间优雅、稳重的生活氛围。

LIFE MOVEMENT

生活乐章

设计公司
邱诚设计

设 计 师
邱振民

项目面积
244m²

主要材料
复古石英砖、镀钛金属板、风化木、铁刀木、铁件、裱布、进口马赛克意大利磁砖、特殊色石英砖

At the hallway, weathered wood in a light color is vividly postured as the protagonist on the walls. Combined with the classical quartz tiles, the design constructs a relaxing and comfortable space foundation. The lengthwise titanium metal board decorates the TV wall and beneath it is the audio-visual machine, whose neat geometric lines and solid cold material degrade the warmth in the living room. Two pairs of couches opposed to each other break the stereotype of the traditional arrangement, and in this way, the artistic design in common area has been transformed into a main power of fashionable atmosphere.

Decorated by the light-colored weathered wood facades, hidden behind the weak or strong lines is the space for storage which can generally store all the family stuff. The red fabric which is a common element in the bedroom decoration is used in the living room, the dining room and the kitchen with the same split method, echoing with the white steel objects to blur the existence of the door leaf leading to private areas. The space with simple outlines is highlighted by the old building patterns on the fabric, enriching the levels of the space, bringing out the owner's casual living attitude and elegant taste of life.

There is plenty natural light in all the space of the project. Passing through the copper metal grid door, you will come into the private area. An open white painted bookshelf goes along the corridor, bringing out the existence of the study area unintentionally.

The floor paved with eye-catching red rust copper tiles stretching to the vertical walls. The magnified black flower pattern on the light blue surface enlarges the guest bathroom space. Bright and colorful camellias used for decoration are the best stage to perform aesthetics.

玄关入口处的浅色风化木，以主角之姿跃然于空间立面之中，在复古石英砖地面的搭配下，构建出闲适自在的空间基础。纵长的镀钛金属板电视墙与下方的视听机盒，利落的几何线条与刚冷的材质中和了客厅里的暖度，而两两相衬的圆弧沙发，也跳出了传统摆设手法的框架，将公共空间的设计艺术，转化成为时尚氛围的凝聚力量。

环绕全屋的浅色风化木立面里，虚实掩映的线条后，藏有可容纳一整屋生活物什的收纳机能。常见用于卧房空间的赭红色裱布，也被设计师以同样的切割手法，与白色铁件错落搭配起来，串连了客厅与餐厨空间，虚化了进入私人空间的门片。布面上斑斓的老建筑图纹，在线条简单的空间中，丰富了空间层次，更带出随性雅痞的生活趣味。

每个空间都拥有充足的自然光照。进入铜金属网门片后的私人空间，沿着廊道设计的白色烤漆开放式书架，设计不着痕迹地呈现出书房空间。

抢眼的赭红色锈铜砖从地面延伸至空间立面，放大线条的黑灰花朵攀附在浅蓝壁面上，放大了客卫浴的空间唯度，而鲜艳多彩的山茶花则是展现空间美学的最佳舞台。

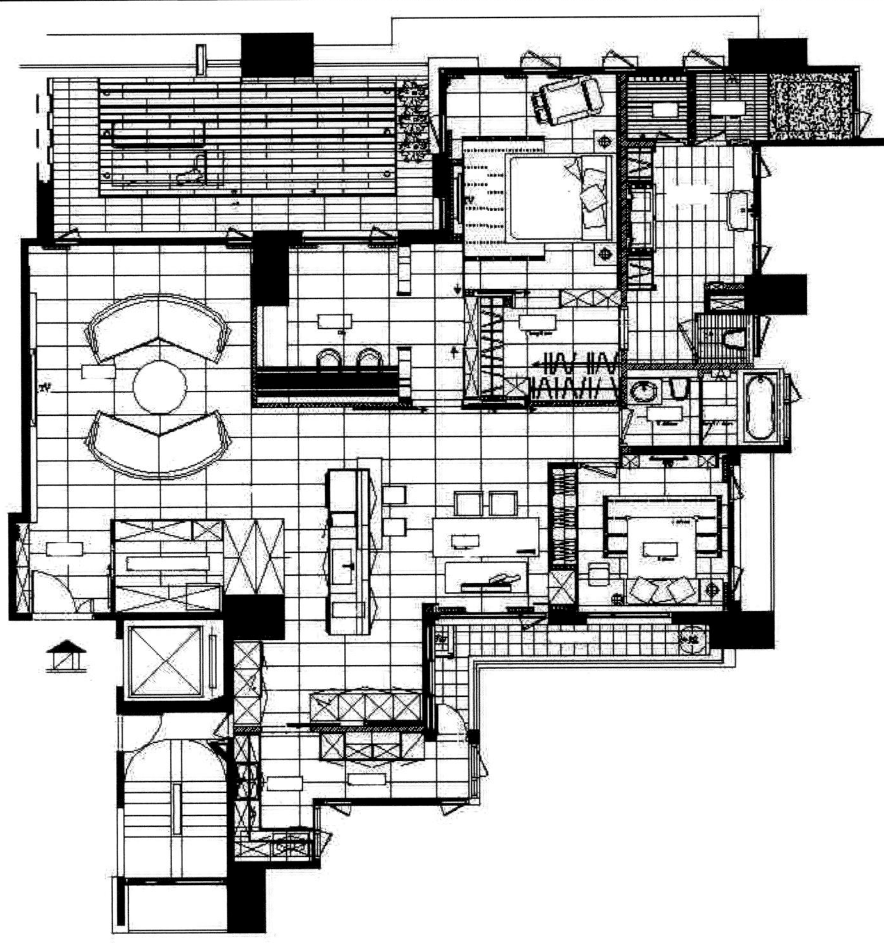

TASTE OF LIFE

极致品味生活

设计公司
黄上科空间设计事务所

设 计 师
黄上科

项目地点
中国台湾

项目面积
$284m^2$

主要材料
钢琴烤漆木皮、进口涂料、石材、铁件、绷布、金箔、皮革、铁件、茶镜、玻璃、木地板、定制家具

According to the owner's request, this project adopts the combination of neo-classical and modern design styles, hoping to create an atmosphere where one can feel every inch of the space thoroughly.

Dark and light colors are used in accordance with collocation of the wall and the furniture. A large area of light colors is applied to the furniture, parts of area in dark colors, forming a sharp contrast. This concise design generates a cozy and refreshing feeling.

The compartment utilizes hollowed hand-made ironware featuring its designing in richness and legerity, totally as a reflection of Art Deco or neo-classical. Oblate iron streaks in a changeable order dividing a horizontal and vertical bar and metaphorically combined cross-streak which indicates the religious belief of the owner's family. The cross-steak elements are used in such space as the screen between the hallway and the dining room, the wall of compartment in the study and the sliding door, wall areas above the bed of master bedroom, and the compartment door of the bathroom. The screen, cabinet and the show stand are framed with bronze-colored gold foil while wooden painted surface uses baked lacquer, which makes the cabinet a dainty and exquisite art as a whole, showing a good taste especially when placing the owner's collection.

The floor design is given varied qualities. The floor at hallway is chequered with stone material surface and shade refracted by brown glass, shaping up a space of reserved glamour and magnificence. Fur design elements are added into some furniture such as desks and cabinets, showing an elegant texture. In terms of the choice of chandeliers, hand-made crystal ones are used for bringing a sense of artistic aura to the whole space.

The combination of every decoration material used in this project reflects its richness and refinement in design concepts, as well as a spirit of modern Art Deco.

按照屋主的意愿，此案设计采用新古典与现代设计语汇相交融的方式，希望营造出可让人回味的空间感受。

在墙面与家具色系的搭配上，深色与浅色交错应用；在家具选择上则以轻浅的家具为主，局部采用深色家具，构筑出反差效果，在形制简练的框架下，给人清爽舒适的感受。

隔间设计运用镂空的手工铁件，结合工艺设计的丰富度与轻巧性，作为 Art Deco 或新古典元素的体现。结构变化的扁铁线条成为垂直水平条分割线，暗含了屋主一家人的宗教信仰十字线条，并将其运用在玄关与餐厅的屏风、书房隔间墙面与拉门装饰、主卧室床头墙面区隔、厕所隔间门等地方。屏风及柜面收边和展示台面特别以古铜色的金箔材质收框，而木皮的漆面也特地使用钢琴烤漆的方式，使柜体如同艺术品一般精致，放上屋主的收藏品后更显品位。

同时，地面上的变化也是极为重要的设计，入门的石材地面加上壁面茶镜的折射，塑造出内敛低调的华丽感。部分家具上如书桌及柜面嵌入了手工皮革，呈现出细致的质感。在吊灯的挑选上，也以手工制作的水晶玻璃来增加整体空间的艺术气质。

各种材质的结合，使此案设计的丰富度与精致度完整体现现代 Art Deco 的精神。

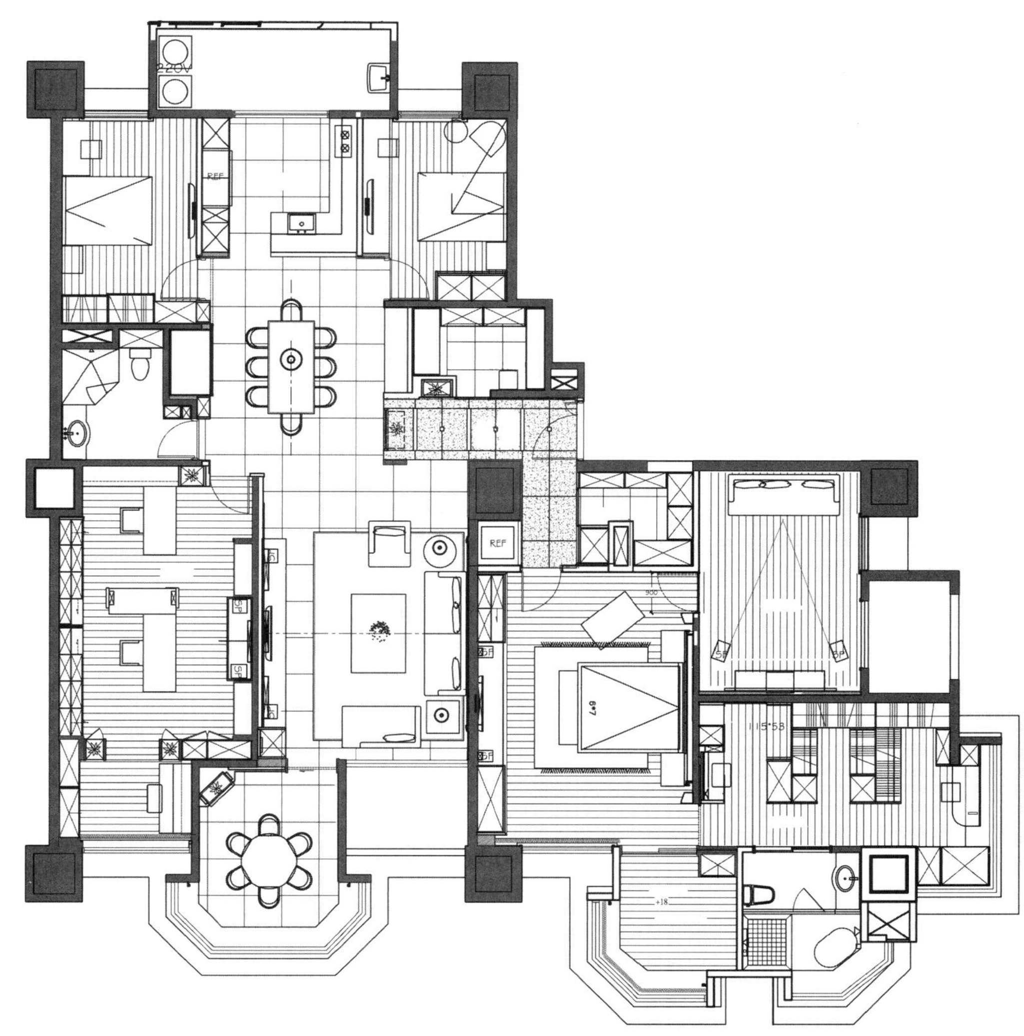

PHOTO ALBUM

NANKANG ELEGANT LUXURIOUS HOUSE

南港优雅豪宅

设计公司
奥迪（国际）室内设计

设 计 师
杜康生

项目地点
中国台湾

项目面积
$264m^2$

主要材料
旧米黄大理石、深金峰大理石、浅金峰大理石、紫檀木地板

The overall atmosphere is built on the accumulation of the humanities and spreads it gradually into every public and private area. Varieties of stone materials interpenetrating in dark colors, give off a warm sense which surrounds the dweller who can feel the inner heart of the space sense.

First Stepping into the living room, one will be attracted by the visual focus created by the crystal on the TV wall. The study adjacent to the living room is vaguely confined by means of semi-perspective effect, like an independent part but in fact interconnected. In space surrounded by glass, individual's view can reach every corner, by which the design of this luxurious house can be upgraded, showing a sense of space from outer to inner.

Walking into the dinner area, one will notice that the decoration elements in living room are extended, presenting a cultured atmosphere. The sliding door confines the dining hall and the kitchen, and the vision is expanded, which echoes with the crystal decoration on the main wall in the living room. Crystal chandelier is placed in the ceiling in the dinning room to create a special atmosphere that make you feel that you are enjoying the dinner in a trendy restaurant.

In the private bedroom, the designer utilizes different hues to zone private and public space by means of using light and warm colors to avoid visual burden caused by heavy ones. Aesthetic effect of light colors used in bedroom builds an elegant, cozy and agreeable design which can settle your inner spirit down. Children's room in the opposite, limited by area, is zoned by low cabinets into sleeping and reading parts, which makes the most use of every inch of space to demonstrate the perfect design.

HOTEL DESIGN
MODERNTOCLASSIC

设计的整体氛围积累了丰厚的人文底蕴，并渐渐地将其移转至每个公、私场所，多种石材的穿插运用，以沉稳的色调延伸，多了一分温暖的色调，环绕着居住者，让空间回归到住宅的心灵层面。

进入客厅，设计便以电视主墙的水晶为端景制造出视觉焦点。邻近客厅区域的书房，利用半透视效果进行空间的模糊界定，看似独立，实则可相互串连。在玻璃围塑的空间之中，人的视野得以拉大至空间的每个角落。同时，此种手法也提升了豪宅的层次设计，由外至内的氛围递层，慢慢地发散出空间的底蕴。

走至用餐区域，客厅中的妆点元素在此延伸，呈现了一种质感，并以拉门界定出餐厅和厨房，再度扩展了视野，同时也呼应了客厅主墙端景处的水晶装饰。餐厅也在天花上方安置了水晶灯饰，让用餐者如同进入到时尚餐厅之中，感受不一样的氛围。

回到私密的卧室空间，设计师利用不同的色调区分出公、私两种空间。采用淡雅的温馨色调，避免使用沉重的色彩，以免出现过多的视觉负担。主卧的浅色美学，反而营造出耳目一新的清新质地，舒适怡人的设计，让心灵能够在此完全沉淀下来。位于另一方的小孩房，由于面积上的限制，采用矮柜作区分，规划出睡眠区和阅读区，让每个空间都能呈现出最完美的设计。

MR ZHONG'S RESIDENCE IN TAIPEI

台北钟公馆

设计公司
天涵空间设计有限公司

设 计 师
杨书林

项目地点
中国台湾

项目面积
120m²

主要材料
钢刷实木板、白色大理石、旋转电视墙、拼花马赛克、LED 灯、木纹装饰板染色、进口木纹磁砖、绷皮造型板、杜邦人造石、茶镜、灰镜、ICI 乳胶漆、金属饰条、定制家具、南方松

摄影师
林福明

The hallway, with double curve ceiling, leads the master into the fan-shaped sitting room by the indirect lamplight. The curve lines of the whole room ceiling show elegant taste and connect the whole space, combined with the curve lines and natural solid board on walls, making the active lines extend between the living room, the dinning room and the public and private areas which enlarge the sense of space.

In the design, lengthened marble countertop table, electric machine cabinet and a transparent glass door served as the separation between spaces are arranged between the dinning room and the kitchen. The wood skin textured door of the kitchen and the wood skin tiles specially used for kitchen floor connect the comfortable natural sense of the entire public space. Paint and steel brush wood boards are specially applied to every door and exhibition space.

The designer designs customized furniture according to the original fan-shaped space in the sitting room so as to build a special arc sitting room which is full of personality. At the same time, the designer makes lighting effect from three directions and applies smooth natural ventilation. Then the public space can be warm in winter and cool in summer without air conditioner. Revolving TV bench made of wood can not only separate the living room and the dining room, but also can be turned towards the dinning room. The designer also planned a hidden world by using southern pine in the balcony, therefore this place becomes a natural garden for male host to plant potted plants.

The design of the entire walls combines beauty and the storage function. The white steel baking closet in the master bedroom makes the space more spacious and comfortable. The hidden bathroom is divided into main bathroom and guest bathroom to make the dry and wet apart, each has its own style and full function. All of the above combined together shows the strong space planning ability of the designer.

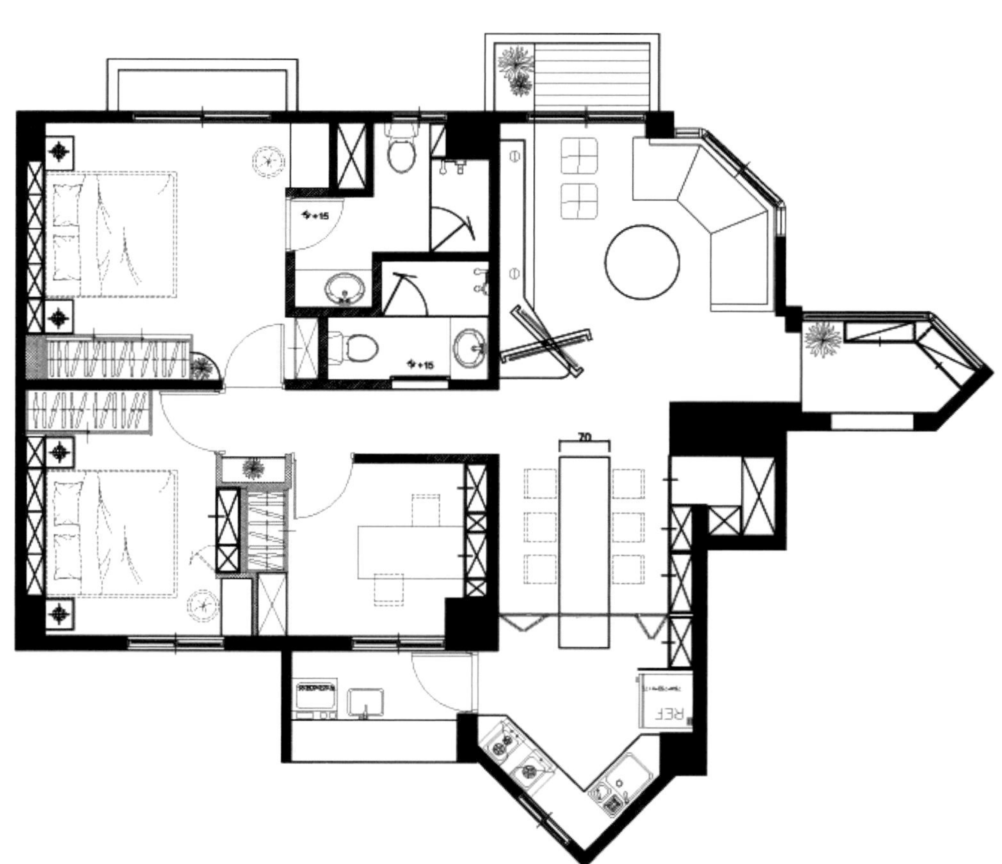
70

玄关设计为双层弧线的天花，利用间接灯光的线条引领屋主进入扇形的客厅空间。全室的天花线条都呈现出优雅的弧线，连接了整个空间。搭配着墙面上共同的天然实木板及曲线，让动线在客厅、餐厅与公私领域之间游走，同时也扩大了空间感。

餐厅与厨房之间设计了加长规格的大理石面餐桌、电器收纳柜、作为空间区隔的透明玻璃门，厨房设计了木纹质感的门片、地板特别选用了木纹材质的磁砖，进而串联起整个公共空间的舒适的自然元素质感，所有的门片及展示空间都特别融合了烤漆与木纹钢刷板的元素。

客厅利用原本的扇形的空间，并定制了家具，营造出别具个性的弧形客厅空间。同时，设计师在格局规划上，营造出了三面采光的效果并能引入自然风，整个公共空间不用空调就能冬暖夏凉。木作的旋转电视柜除了是客厅和餐厅的空间区隔外，还可以转向餐厅。阳台利用南方松规划出一个隐蔽的小天地，作为男主人种植盆栽的自然园地。

全室的墙面设计除兼顾美观外，更规划了丰富的收纳功能。主卧房的白色钢烤衣橱让空间看起来更加宽敞舒适。隐藏的卫浴空间，设计师利用不规则的原始空间变化出干湿分离的主客卫浴，两间各有不同风格，同时还兼顾完整的机能，因此更显现出设计师强大的空间规划能力。

MR PENG'S RESIDENCE IN TAICHUNG

台中彭宅

设计公司
大言室内装修有限公司

设 计 师
黄金旭

参与设计
廖怡菁

项目地点
中国台湾

项目面积
330m²

主要材料
大理石、水性木器漆、实木皮

摄影师
刘俊杰

"Simple Life" is the proprietors' life proverb. So this case is in needs of environmental protection, energy saving and low profile to give effect to a healthy, simple, unadorned life experience and real life functional demands of the owners.

The design takes environmental protection material as the main option. Entering the sitting room, the white TV marble wall separates the living room and the study. On the other side, we put the import audio TV ark bought by the owner into the TV wall to make the whole living room clean, agile, building a low profile and simple atmosphere. Both sides of the living room and the dining room corridor walls are designed as real wood with environmental paint forming natural feelings in the space. The ceiling lamps in the living room and the dining room are all LED lights combined with furniture materials, colors and pure white walls, building a entire case spirit.

The representation of bedrooms depends on the color and the functions the family members' favor, and the change of color and the bedrooms' materials echo with furniture styles to further represent different space styles.

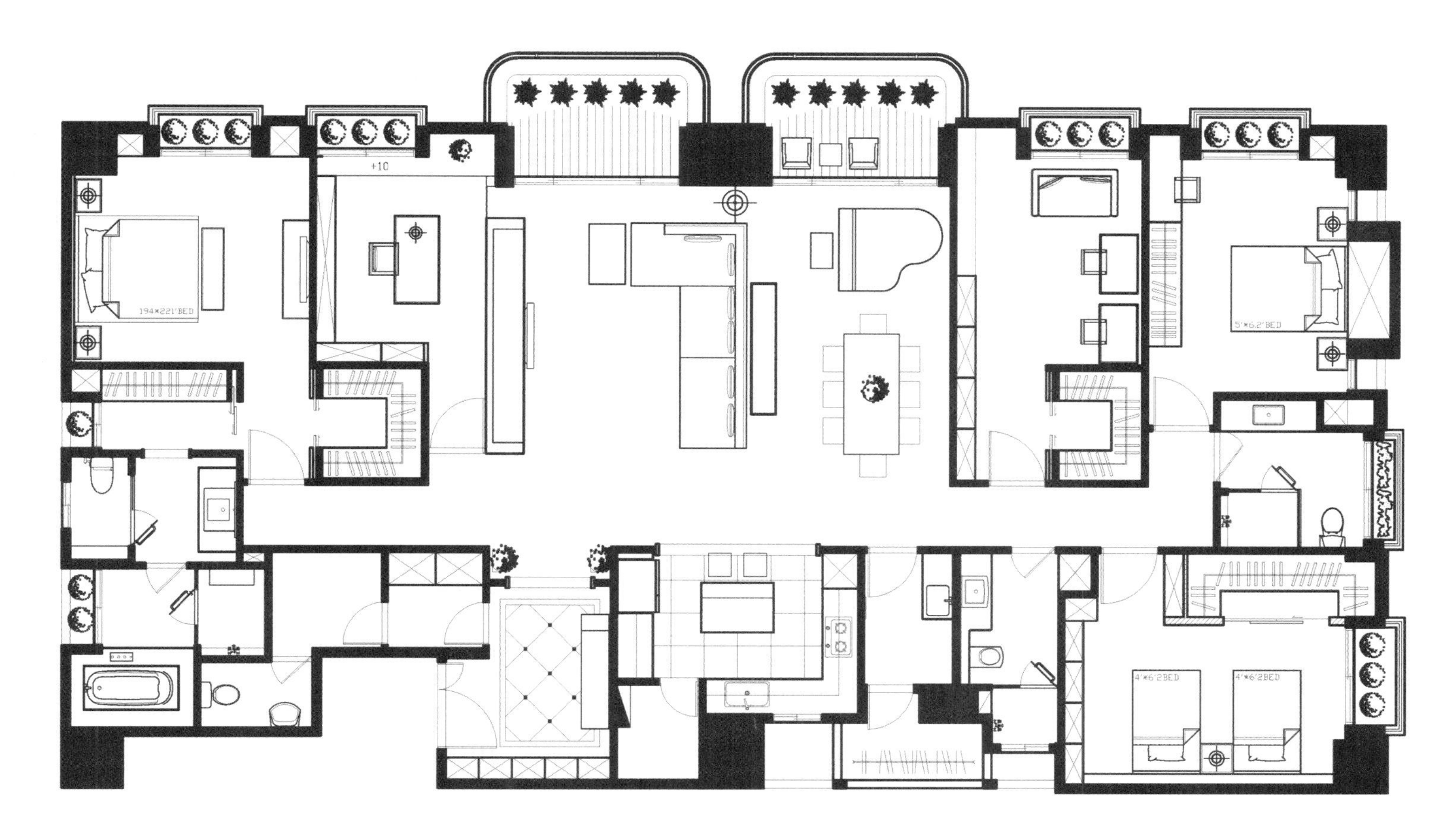

“简约生活”是屋主奉行的生活格言，因此本案以环保、节能、低调为出发点，以落实屋主健康、简单、朴实的生活体验和实际生活机能需求为导向。

全案设计主要材料皆以环保材质为主要选项。进入客厅，电视墙面以洁白、素净的大理石墙界定出客厅与书房。另一方面，屋主特别选购的进口音响电视柜被嵌入电视墙内，使整个客厅干净、利落，营造出低调、简单的氛围。厅客、餐厅两侧廊道的墙面设计上采用实木纹理材料，面刷环保漆，在空间中形成自然语汇，客厅、餐厅天花板灯具全部用LED投射灯，经过整合的家具材料、颜色及纯粹的白墙，共同塑造出全案的设计精神。

在卧室空间表现上，设计依据各个家庭成员的喜好选择色调及生活使用机能。色彩的变化与各个卧室空间材料的表现形式都呼应了家具的式样，进而呈现出空间的不同风貌。

SIMPLE NATURE
简约本质

设计公司
金湛设计

设 计 师
凌志谟

Brief vogue, functional combination

In modern design, brief or simple style is not easy to accomplish for an actual living space where precise space design rules are hard to maintain, so fashionable and Northern European elements added into this Minimalist project can make the whole space no longer monotonous, but more lively, vivid and lifelike.

Self-evidence of every area created by intermediary space of the actual and fake design:

Partition shelves whose cantilever overhangs serve as an intermediary element between the open space and the kitchen, making it possible for a traditional closed kitchen to be open-ended. So, the center of family activity is established by integrating the kitchen and the indoor space. Additionally, the other half of the partition shelves becomes the space of the master's study and behind the plank wall with the large door leaf is the space of his or her bedroom. Isolated space is formed easily so as to keep privacy when shutting the plank.

There are two doors in the entrance of the bedroom: an actual (wooden) door and a fake door – steel glass door, both of which define different space ranges which are not fixed any longer and therefore overlapped space becomes private as well as open-ended.

The whole space reduces the active line to the lowest state and visualizes the only active line and makes it as straight and long as possible to highlight the entrance and to display end wall interpreting the space atmosphere. The impending cupboard separates the active line from the dining room without dividing the whole space, and it adds the sense of penetrating without cutting off the whole space into pieces.

THE DESIGN HOTELS
BOOK

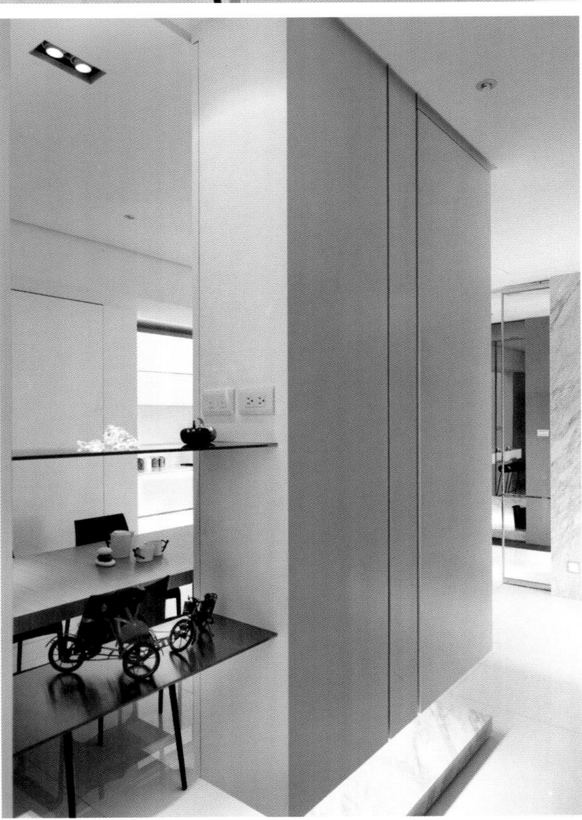

极简时尚，机能并融

现代主义中的“极简”在每个真实的居住空间中都难以实现，过于严谨的空间规则让人们难以维持。所以在极简主义之下加入时尚元素与北欧元素，可使空间不再单调，也更显活泼，当然也就更贴近生活。

虚实中介空间手法创造各单元空间的自明性：

将传统封闭式的厨房尽可能地变更为开放式，悬臂出挑的中岛为开放空间与厨房的中介元素，不但让厨房与内部空间相结合，更创造出家庭的活动核心。另外，中岛的后半部是主卧的书房空间、大扇的门片墙后是主卧空间，门关起后就能轻易隔绝，保持私密性。

主卧室的入口有两扇门，一是实体（木作门）一是虚体（不锈钢玻璃门），两个门界定了不同范围的空间，空间领域不再固定，重迭的空间变得既私密又开放。

整户的空间动线减到最低的状态，并且将唯一的动线可视化，尽可能拉直拉长，强化进口的效果，并由端景揭示空间氛围。悬空柜分开了动线与餐厅，却不隔断空间感，悬浮的柜体更增加了穿透感，空间不会被切割得过于零散。

TASTE THE LIFE

品生活

设计公司
TBDC 台北基础设计中心

设 计 师
黄鹏霖、黄怀德

参与设计
林容宇

项目地点
中国台湾

项目面积
290m²

主要材料
烧毛面石英砖、烧毛面花岗石、钢刷橡木、毛丝面不锈钢、铁件烤漆、烤漆玻璃、黑色玻璃、环保木地板、环保涂料

摄影师
王基守

The owner of the residence loves red wine, so the design of the space has a gorgeous and meaningful disposition. Life in it is like to have a taste of mellow red wine, and you can enjoy the pleasure of life by watching its color, smelling its fragrance, and tasting its flavor.

The whole space is divided into different parts according to the functions, including the hallway, the living room, the study, the dining room, the working area, the kitchen, the wine cellar, the working balcony, the master bedroom, the master bathroom, the changing room, the children's room, the guest room and the guest bathroom, etc. Except the parts in which privacy are in high priority, other parts of the space are fully open to permit natural light in and to make every corner in the space bright.

No redundant design is used here and the exquisite effect is achieved by the proportioning of materials and the utilizing of illumination. Penetrating materials spread randomly on the achromatic color background, and blocks and planes made of natural timber balance the space perfectly. Industrial delicacy and natural texture are coordinated together to create a unique effect. Lights are arranged in different positions and angles to meet the needs of different surroundings. All the design makes the space meaningful and reserved, and meanwhile, the unique taste and noble disposition are fully presented, like well-structured red wine with abundant and balanced flavor, high in quality and fresh in taste.

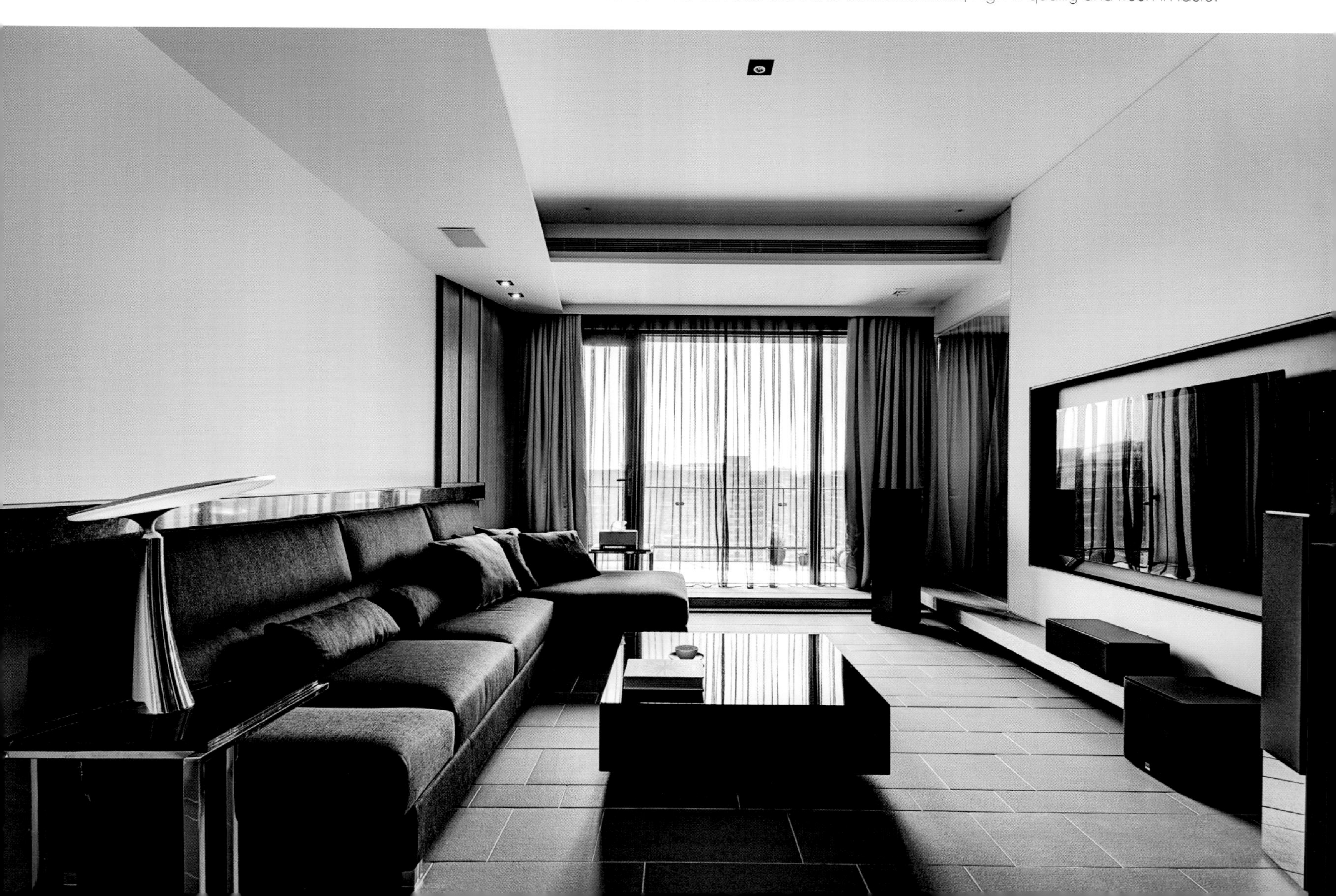

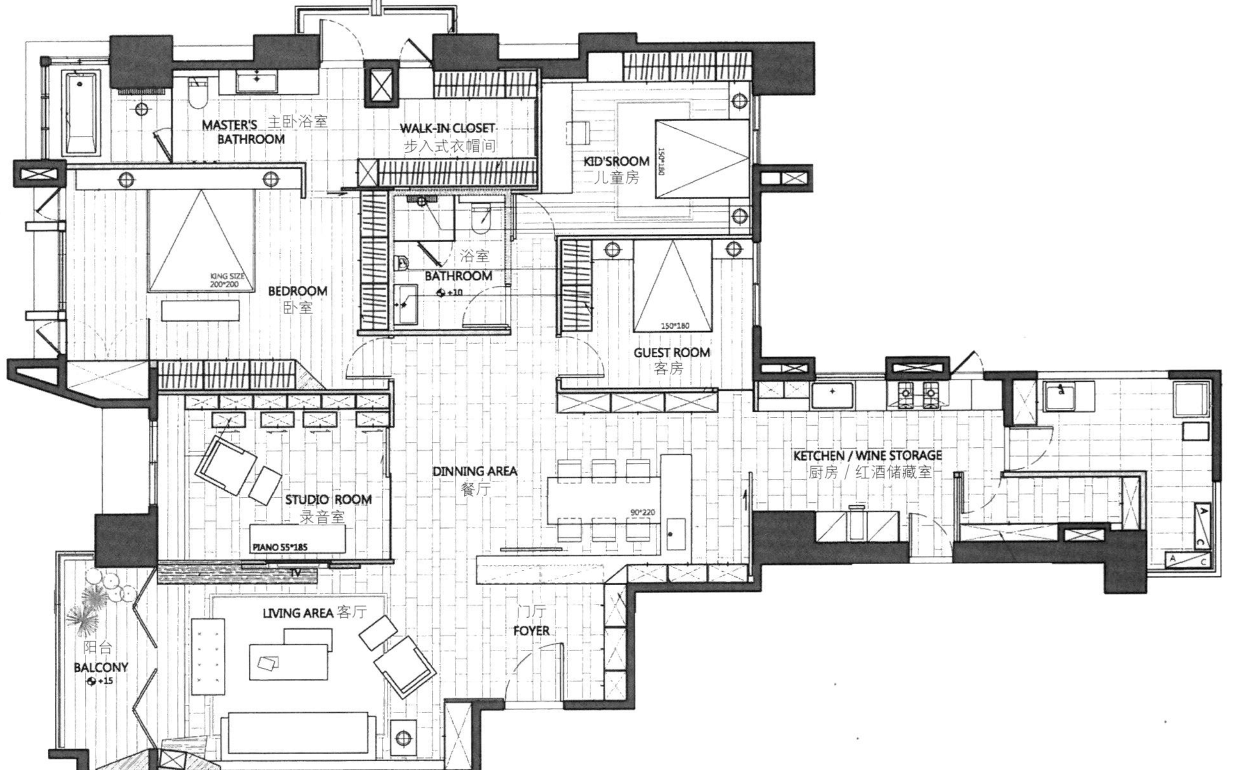

此案的屋主非常喜欢红酒，因此设计师尝试赋予空间一种简单又富有意蕴的气质，如同品酒，观其色泽，闻其香味，品尝啜饮，从中感受人生的愉悦。

空间因使用需要，切割为玄关、客厅、书房、餐厅、工作吧台、厨房、酒窖、工作阳台、主卧室、主浴室、更衣室、儿童房、客房及客用浴室等使用空间。除了私密需要，其余空间均希望能充分地开放，让自然光能最大程度地延伸至空间的每个角落。

没有多余的主观的造型，设计通过材料的配比与灯光的运用来处理空间。以无彩色的基调为底，穿透性的材料错落其中，天然木料在空间内以板块与量体的形式呈现，平衡了空间的轻重关系，呈现了工业的精致与天然肌理的触感。同时，应情境需要做了不同角度与方位的光源配置，让空间在内敛的底蕴中，挥洒出独特的风味与高贵的气质，如同红酒丰富而均衡、质优而新鲜的口感一般。

KNIGHT

GIORGIO ARMANI

LOUIS VUITTON

NATIONAL GALLERY BUILDING B

国家美学馆 B 栋

设计公司
台湾禾林室内设计有限公司

设 计 师
吴瑞麒

项目地点
中国台湾

项目面积
350m^2

主要材料
石材、金箔、烤漆板、壁纸、铁件工艺

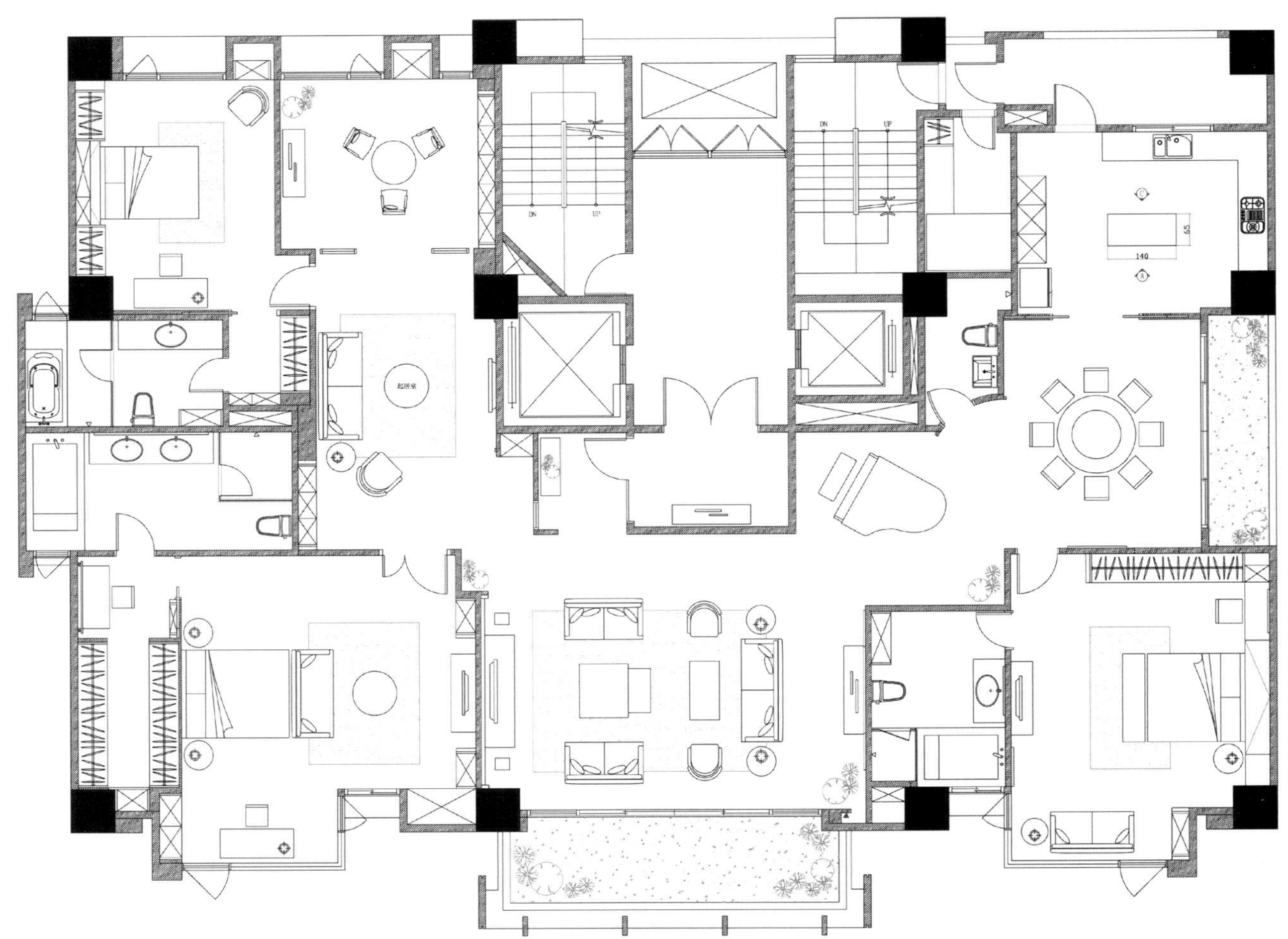
DN
UP
起居室
140
65

In the hallway, three-dimensional golden and silver foil echoes with the wide rim of the featured shoes cabinet, combined with the sculpture symbolized "A Happy Ending", creates a typical mention with luxurious style. Potted flowers are placed in accurate positions, which echo with the view outside the window.

The space is designed in a stable tone, with indirect lines. From the straight yacca in the balcony to the sofa in the sitting room, the space is arranged grandly in the order of front, center and main area. The marble, like a landscape painting, fixed on to the main wall, decreases the stiff and hard look of the wall. The silver foil applied in a proper portion adds variety to the chosen materials.

Following the circulation and pass through the piano room, it is the dining area. The ceiling here with a simple outline, collocates with crystal main light to create a harmonious dinner atmosphere enriched with sweet melodies. Through a glass door which works as the bridge to the kitchen, it is apparently to see tawny diamond mirror on both sides of the main symmetric line, besides, the bracket lamp with dark cover and the three-dimensional square ceiling perfect the composition of a "square" and a "circle" between the kitchen and the dining room. On the other side of the space are the sitting room and the study. The featured screen optimizes the functional positions, and large tawny mirror is embedded into the wall behind the sofa, which visually enlarges the space and improves the quiet and harmonious atmosphere.

The big main bedroom is designed with double doors, the main wall is decorated with elegant wall paper implying the client's taste, and all the furniture, decorations and layout are presented in an almost perfect way. The guest room beside the study is decorated with Japanese style embossed wall paper, revealing an extraordinary style. The minor master bedroom behind an invisible door in the dining room not only provides adequate storing space, but creates a calm and elegant atmosphere which makes people reluctant to depart.

入门处的玄关空间，墙上立体的金银箔与端景处鞋柜的宽边相搭配，与象征“大圆满”的雕塑品一同建构出豪宅的气势。设计恰切地摆放了落地盆栽，与窗外景色相呼应。

客厅以沉稳的色调、间接的线条，从阳台上直挺的罗汉松，一直延续到客厅的沙发，以前、中、主位区有序的安排，呈现出大器的格局。如山水画的大理石轻拂僵硬的主墙，比例恰当的银箔使材质富于变化。

循着动线，经过琴房来到餐厅，线条简单的圆顶天花，搭配主灯上的透明水晶，琴音淡淡丰富了餐叙情景。与厨房间的联系，推开玻璃门，就能看到对称主轴左右两侧的茶色菱镜与深色的灯罩壁灯，立体的方顶天花，这些都完美展现了其与餐厅之间“方”“圆”的气质与风采。空间的另一端，卧房前是起居、书房空间，选用造型隔屏，优化了空间定位；大面积的落地茶镜静静地嵌入沙发墙中，以其反射功能放大了空间尺度与静谧氛围。

大面积的主卧具备双开大门，淡雅的主墙壁纸烘托出屋主的品位，摆设、规划以近乎完美的要求呈现出来。书房旁的次卧以日式立体压纹壁纸凸显超凡气质。位于餐厅暗门内的次主卧中，功能到位的收纳空间，满室的沉稳素雅，令人驻足难忘。

索引

INDEX

宽月空间创意

吴奉文、戴绮芬

2009 年第七届现代装饰国际传媒奖“年度最具潜力设计师大奖”；

2010 年第八届“现代装饰国际传媒奖”“年度家居空间大奖”；

2010 年 IAI AWARDS 亚太室内设计双年大奖赛“最佳住宅空间设计大奖”；

2011 年第九届现代装饰国际传媒奖“年度空间配饰大奖”；

2011 年第四届全国十佳配饰设计师；

2011 年 IAI AWARDS 绿色设计全球大奖“最佳家居空间大奖”；

2012 年“2011 年香港全球设计协会住宅空间设计奖”。

公司网址：http://www.lunarian.com.tw

常玉国际设计工程有限公司

吴衍霖、许秀娜

设计是为呈现更美好的生活空间与方式，创造、寻找出每个人与空间互动的独一无二的特质。

公司网址：http://www.as-tw.com.tw

天涵空间设计有限公司

杨书林

设计师相信生活即是艺术，空间设计的理想就是以人为本，以丰富的空间设计经验及建筑学背景，聆听与整合客户的需求，在美感与实用间取得平衡，为每一位客户都创造出梦想与个性兼具的专属空间设计。

公司网址：http://www.skydesign.com.tw

玳尔室内设计有限公司

朱志峰

1998 年成立玳尔室内设计公司，专攻日式禅风，并于 2006 年对于住宅之风格提出“疗愈系”的路线，作品中呈现出商业空间的“创意拔群，思路奔放”与住宅空间的“以人为本，坚持素朴”。将两种看似冲突的思维却完美融合在个案中，因独树一帜之细腻感性风情，荣获网络票选最具风格设计师！ 而“疗愈系空间”一词，已成为业界风格定位上的专业用词。

公司网址：https://www.facebook.com/pages/玳尔室内设计有限公司/301243963260903

近境制作

唐忠汉

CSID 中华民国室内协会 理事

中国科技大学室内设计课程客座讲师

2007 年 TID 台湾室内设计大奖居住空间大奖；

2008 年 TID 台湾室内设计大奖居住空间入围；

2009 年 TID 台湾室内设计大奖居住空间大奖；

2009 年 TID 台湾室内设计大奖商业空间大奖；

2009 年 TID 台湾室内设计大奖居住空间入围；

2010 年 TID 台湾室内设计大奖居住空间金奖、居住空间大奖；

2010 年 TID 台湾室内设计大奖商业空间入围；

2010 年好宅配大金设计大赏；

2011 年 TID 台湾室内设计大奖居住空间、商业空间大奖；

2011 年 TID 台湾室内设计大奖居住空间入围奖、家具入围奖；

2011 年国际空间设计大奖“艾特奖”最佳住宅空间、最佳商业空间设计入围奖；

2011 年金外滩最佳创新设计奖、最佳办公空间奖、最佳居住空间奖

公司网址：http://www.da-interior.com

奥迪（国际）室内设计

杜康生

奥迪的核心价值是扎实的设计理论和设计风格的拿捏与掌握。从繁复的古典到现代的简洁，奥迪都能展现其精髓。面对所有个案，均以顶级材质搭配细腻作工，在比例的拿捏、美学的概念、材料的运用、色彩的着墨上兢兢业业求其完美。

公司网址：http://www.audi-design.com.tw

玉马门创意设计有限公司

林厚进

设计师认为，基地条件、居住者的喜好、个性及生活形态，是主导空间成形的关键。透过设计师的创意美学衍生出的居家风格应该超脱文字定义的范畴，不拘泥于形式，藉由点、线、面实际串连构成的空间表情来自我表述与定义，这样的设计风格，才能更丰富而精采!

公司网址：http://www.yuma-m.com

动象国际室内装修有限公司

谭精忠

2009 年被新浪地产易居地产研究院评选为中国房地产室内最佳机构；

2009 年入选中国环境设计年鉴 2009；

2010 年入选中国环境设计年鉴 2010；

2010 年中国 TOP10 办公空间照明应用设计最佳人文气质奖。

公司网址：http://www.trendy-interior.com

怀特设计

林志隆

设计的角度，抓的是层面的变化与质感，怀特设计从与业主的沟通开始着手，以个性为铺陈、品位为架构，捏塑出以人为主、感受性为辅的空间思维。

设计，是有想法、有故事性的，整合生活情绪发展成为空间的情感，延续出丰富的生命力。

一个新的材质或是材质的变化，这些都代表着设计本身的价值创造，透过这些细节与表现语汇所累积起来的品牌印象，成为公司希望创造的一眼便能看出的怀特式风格。

公司网址：http://www.white-interior.com

珥本室内装修设计工程有限公司

陈建佑、曾耀征

室内设计不仅是满足设计师对美感的期待，还是面对生活的一种态度。要将客户的需求与对基地的分析相结合，具体内容包括良好的动线机能规划、材质与光线的演绎、造型分割的比例、计划性的照明，还包括家具摆饰的挑选以及搭配空间的形象设计，要为客户提出中肯的建议与体贴的服务，因此，一个项目的最终成果是具体性及施工技术完整的呈现。

公司网址：http://www.urbane.com.tw

邱诚设计

邱振民

运用各种空间美学提升空间质感，无论是奢华的新古典，还是现代时尚的简约优雅。在进行空间规划的同时，更应执着于空间质感的营造，以打造舒适的生活空间。

公司网址：http://www.chi-cheng.com.tw

异国设计

王文亚

以异国作为公司名称，源自公司的英文名 Exotic，即“异国风情”之意；撷取各国多样性元素，融入设计师的专业与用心，创造出异国设计独树一帜的风格。公司在设计之初，即秉持对设计的执着，针对每一个案均投入专业团队，充分展现出异国设计的多样性、变化性、包容力及不受局限的设计理念。

公司网址：http://www.exotica.tw

大言室内装修有限公司

黄金旭

中原大学室内设计研究所硕士
岭东科技大学科技商品设计系兼任讲师
2009 年台湾室内设计大奖商业空间类入围；
2010 年台湾室内设计大奖居住空间类 / 单层 TID 奖；
2011 年台湾室内设计大奖工作空间类 TID 奖；
2012 年金创奖商业空间类金奖。

公司网址：http://www.greatword.com.tw

界阳 & 大司室内设计

马健凯

一直以来，界阳设计以秉持黑白、前卫、时尚为设计理念，借由玻璃、石材与金属材质，搭配现代科技光照，打造出一场场精彩绝伦的设计盛宴。

为了服务不同的业主，界阳设计打造出其第二品牌——大司设计，主要以自然、人文、典雅为设计理念，借由木皮、石材、自然光、绿意等自然元素，为空间注入自然人文的概念。

公司网址：http://www.jie-yang.net

黄上科空间设计事务所

黄上科

于中原大学建筑系、淡江大学建筑研究所、兰阳技术学院建筑系兼任讲师

设计师秉持“少即是多”的设计理念，将实用机能藏于精巧的设计之中，搭配适当的家具，凸显空间的美感。

公司网址：http://www.skh-design.com

Danny Cheng Interiors Ltd.

郑炳坤

毕业于加拿大多伦多 International Academy of Merchandising and Design，于 1996 年回港与合伙人成立公司。2002 年自立门户，开设 Danny Cheng Interiors Ltd.，以设计住宅、示范单位及商业空间为主。多年来，其简约、富建筑感及空间感的设计曾获得多个香港乃至世界性的大小奖项共七十余个。

公司网址：http://www.dannycheng.com.hk

台湾禾林室内设计有限公司

吴瑞麒

铭传大学商业设计系
中原大学室内设计研究所
铭传商设系校友会会长
台湾室内设计协会理事

公司网址：http://www.holin.net.tw

金湛设计

凌志谟

面对设计，凌志谟的态度是谦卑的，将美学素养视为设计者的本分，不将自己的想法妄加于空间之中，而是尊重业主对于“家”的期待，以人与家的关系为主题，创造真实中有梦想的主题故事。

曾经面对业主推翻原创又再次接受原创的冗长讨论，凌志谟深信“业主才是空间主人”的理念，他愿意等待，给予对方充分的思考时间。他自信地说：“设计师是从整体去看美感与功能，总是比业主预先完成思考，因此业主最后都会尊重专业，也会对最终结果满意。”的确，只要多一份敏感，多思索住宅的使用功能，多为生活的舒适性着想，业主一定会感受到这份专业与贴心，欣然接受设计师的精心创意。一颗谦虚的心、一份拥有创意的自信，使凌志谟的作品能够于风格中准确地抓住美感与精髓。设计在他的手中变得有趣，变得精致，变得拥有无限可能，精彩的生活于此开始。

公司网址：http://www.goldesign.com.tw

TBDC 台北基础设计中心

黄鹏霖、黄怀德

德国 iF design 台湾地区会员
CSID 设计协会会员
2011 金外滩最佳居住空间优秀奖、最佳色彩运用优秀奖；
中国 TOP10 办公空间照明应用设计年度人物奖；
德国 Red dot design award 2007 商业及公共空间类大奖。

公司网址：http://www.asia-bdc.com

城市设计

陈连武

淡江大学建筑研究所建筑工程硕士

2009 年大金设计大赏佳作奖;

2009 年中华民国杰出室内设计作品金创奖银牌奖;

2010-2011 年海峡两岸四地室内设计大奖住宅类金牌奖;

2010 年亚太室内设计双年大奖赛新锐设计师奖;

2010 年亚太室内设计双年大奖赛优秀作品入选;

2010 年大金设计大赏铜牌奖;

2011 年金堂奖年度优秀作品奖;

2011 年金堂奖年度十佳住宅公寓空间作品奖;

2011 年 IAI AWARDS 最佳家居空间设计大奖;

2011 年 IAI AWARDS 最佳企业空间设计大奖。

公司网址：http://www.chainsinterior.com

德坚设计

陈德坚

德坚设计（Kinney Chan & Associates）于 1995 年成立，为酒店、酒吧、餐厅、住宅、商业店铺和企业办公室等多个行业提供广泛而多元化的室内设计和项目管理服务。

一直以来“创造力和原创性”是鞭策 KCA 不断锐意求新的动力来源。KCA 的设计师团队都是勇于尝试和充满革新意识的，他们视室内设计为一种真正的艺术形式，而不仅仅是室内和室外的空间规划和物料的配搭。

公司网址：http://www.kca.com.hk

台湾建构线设计有限公司

沈志忠

设计理念：

尽可能让艺术机能化，空间艺术化，在机能范围内发挥最大的创意价值，让自已的想法像做纯艺术般尽情发挥，不被局限。

获奖纪录：

2008 年 IAIC 亚太室内设计住宅空间铜奖;

2008 年 TID 台湾室内设计大奖新趋势设计金奖;

2009 年 IAI 亚太室内设计住宅空间铜奖;

2010 年 APSDA 亚太空间设计师协会设计大奖;

2011 年 TID 台湾室内设计金奖;

2011 年英国 FX 设计大奖入围总决选（全球入选 10 位）;

2011 年德国 IF 设计奖传达设计奖（全球 256 国家中脱颖而出）。

公司网址：http://www.x-linedesign.com

玮奕国际设计工程有限公司

方信原

1998 年创立玮奕设计;

1999-2000 年于欧洲地区游学;

2004 年于美东纽约设计事务所实习;

2005 年正式创立“Newspaper”品牌。

公司网址：http://www.lw-id.com

陆希杰设计事业有限公司

陆希杰

1989 年毕业于东海大学建筑系，1993 年取得英国 AA 建筑联盟硕士学位，在英国期间曾于 Raoul Bunschoten 事务所担任设计师，而后回国成立陆希杰设计事业有限公司，从事建筑及室内设计、家具设计、产品设计等相关研究开发工作。

公司网址：http://www.shi-chieh-lu.com

杨焕生设计师事务所

杨焕生、郭士豪

杨焕生

东海大学建筑设计研究所硕士

杨焕生建筑 / 室内设计事务所 执行总监 & 主持人

郭士豪

云林科技大学

杨焕生建筑 / 室内设计事务所 设计总监 & 协同主持人

2009 年 TID 设计大奖“商业空间类”入围;

2009 年 TID 设计大奖“工作空间类”入围;

2011 年 TID 设计大奖“居住空间类”入围;

2011 年 TID 设计大奖“临时建筑类”入围;

2011 年 TOP DESIGN 顶尖设计奖;

2011 年两岸三地新锐设计奖;

2012 年金外滩设计奖。

公司网址：http://www.yhsstudio.com

后记

本书的编写离不开各位设计师和摄影师的帮助，正是有了他们专业而负责的工作态度，才有了本书的顺利出版。参与本书的编写人员有：

吴奉文、戴绮芬、朱志峰、林厚进、陈建佑、曾耀征、吴衍霖、许秀娜、唐忠汉、谭精忠、邱振民、吴衍霖、许秀娜、杨书林、杜康生、林志隆、王文亚、邱振民、马健凯、黄上科、郑炳坤、吴瑞麒、凌志谟、黄鹏霖、黄怀德、陈连武、陈德坚、沈志忠、方信原、陆希杰、杨焕生、郭士豪、王真、曹亮、王丽娟、卢晓娟、王丹、何心、钱源、张华慧、杨灿灿、王超、林奕佳、张方展、刘洋、何雪君、李青云、邱凌云、刘健琨、孙璐佳、董喆、许雅杰、张丽芳、赖小珍、邹筠婷、王清、李嘉欣、梁家玉、钟舒欣、黄子平、金敏、吴坤燕、孙林林、魏越、张琳、罗敏、郑婉媛、肖丹、张弛、邓颖、胡志敏、戴乐玲、陈勇妮、姜艳

ACKNOWLEDGEMENTS

We would like to thank everyone involved in the production of this book, especially all the artists, designers, architects and photographers for their kind permission to publish their works. We are also very grateful to many other people whose names do not appear on the credits but who provided assistance and support. We highly appreciate the contribution of images, ideas, and concepts and thank them for allowing their creativity to be shared with readers around the world.